OF MEN
AND NUMBERS

THE STORY OF THE GREAT MATHEMATICIANS

JANE MUIR

DOVER PUBLICATIONS, INC.
New York

In memory of
Sally Henry Kitchen and Gigi Gottlieb

I should like to take this opportunity to thank the many people who helped me in one way or another to write this book. Specifically, I should like to thank my husband for being so very patient and encouraging and for helping type the manuscript. I should also like to thank Theodore Price for his invaluable criticisms; Mrs. George W. Henry for her suggestions and the loan of much research material; and my mother, Helen Muir, for her assistance in translating several important source books.

Bibliographical Note

This Dover edition, first published in 1996, is an unabridged, unaltered republication of the work originally published by Dodd, Mead & Company, New York, 1961.

Library of Congress Cataloging-in-Publication Data

Muir, Jane.
    Of men and numbers : the story of the great mathematicians / Jane Muir.
       p.   cm.
    Originally published: New York : Dodd, Mead, 1961.
    Includes bibliographical references and index.
    ISBN 0-486-28973-7 (pbk.)
    1. Mathematicians—Biography.   I. Title.
QA28.B8   1996
510′.92′2—dc20
                                        95–39752
                                           CIP

Manufactured in the United States of America
Dover Publications, Inc., 31 East 2nd Street, Mineola, N.Y. 11501

Contents

Pythagoras

582 B.C.(?)-507 B.C.(?)

Euclid

C.300 B.C.

Archimedes

287 B.C.-212 B.C.

The names of the men who pioneered in mathematics are lost in the same ancient mists that obscure all of mankind's early history. There are no records of the cavemen who, eons ago, first conceived the idea of counting: 1, 2, 3, 4—many. As language evolved, so did counting and very simple arithmetic. Men added 2 arrows to 3 arrows and got 5. Often several sets of numbers were used, depending on the objects being counted. Vestiges of these multi-number systems occur even in modern usage where a couple, a pair, a brace, and a duo all mean "2."

Millennia went by. Hundreds of generations rose from and sank back into the earth before the spoken number evolved into a written one. At first the notation system consisted simply of pictures of each object being counted—three bison painted on a cave wall recorded the number of animals a hunter had killed; four arrows scratched into a piece of bark indicated the quantity of his weapons. Even the early Egyptians used pictures of objects to show "how many." Neither the spoken nor the written number systems were as yet abstract, although pictures were

1

the first step in that direction, for a picture itself is not the actual object but only a symbol of it. Yet for each object being counted, a corresponding number of pictures were used. Numbers could not yet be thought of apart from the objects or pictures thereof being counted. Number was not a separate entity.

With the rise of personal property—domestic animals, land, bushels of wheat, etc.—mathematics took a giant stride forward. Property had to be taxed; therefore, the chief or leader or king had to know how much property each of his subjects had in order to levy taxes. A picture of each and every object being counted was too cumbersome and systems of number notation developed —notches on a stick, lines and dots pressed into a clay tablet. Instead of using several sets of symbols, one for cows, another for bushels of wheat, one set was used for all objects and the symbols were given names: one, two, three, four, five, etc. Numbers now had an entity of their own, completely divorced from the things being counted.

This abstraction of numbers was the beginning of mathematics, for men could now deal with numbers or symbols rather than actual objects. Addition and subtraction could be carried out without physically increasing or diminishing a quantity of objects.

Still the number systems were not perfect. The same symbol was often used for more than one number; it could mean five or twenty or one hundred, and its value had to be guessed from its position in the series—a state that persisted well into the Middle Ages. Multiplication and division required hours to perform and could only be done by experts. Try dividing MCCIX by LVI.

But history could not wait for the number system to be perfected. Mathematics advanced in spite of its poor notation—and advanced brilliantly, with an achievement that five thousand years later is still one of the wonders of the world.

This near miracle was the building of the giant pyramids in Egypt. More than memorials to dead kings, they are monuments of mathematical triumph. The angles at the bases of the pyramids had to be exactly 90° or the sides would not meet at one point at the top. Each side had to slope inward at the same rate. Each stone had to be placed precisely. The temple of Amon-Ra at Karnak was built so that on the longest day of the year the setting sun shines directly through the building, from the front door to the back wall. Obviously, this was no hit-or-miss affair. Yet it was all done with crude measuring devices by men whose number system was so clumsy that a simple problem in multiplication required the services of a professional mathematician.

The Egyptians are among the first known peoples to utilize the basic rules of geometry. On the American hemisphere, the Mayans made a start, too, as evidenced by their remarkable temples; but their civilization fell without their discoveries ever entering the mainstream of man's learning. Credit must go instead to the Egyptians for taking the first crude steps with such glorious results.

The Egyptians were impelled to learn something about geometry, not from intellectual curiosity but for practical reasons. Every year the Nile overflowed, wiping out landmarks and boundaries, and, as every schoolchild knows, spreading fertile silt along its banks. In order to restore the boundaries, the Egyptians had to learn to measure the land. Therefore, they turned to geometry. The very word in Greek and Latin means "to measure" (*metrein* and *metria*) "the earth" (*ge* and *geo*).

Geometry never achieved the status of a science with the Egyptians. It never was anything more than a collection of rules and rough measurements arrived at by trial and error—the circumference of a circle, for instance, was calculated to be $(\frac{4}{3})^4$ of the diameter, or about 3.1605.

That the Egyptians did not go further in mathematics is not surprising. A highly religious people, their lives were oriented toward death rather than life. They built their tombs and their ships for the dead—everything was directed toward the life hereafter; nothing mattered in the present one, certainly not mathematics. Furthermore, the only educated people were the priests, who kept all learning to themselves. Once they had figured out how to tell from the position of the stars when the Nile would overflow, they dropped their study of astronomy. Nor did they make any effort to pass along their knowledge to others, for it was to the priests' benefit to keep the people ignorant and dependent on them. Learning, as a result, never flourished.

Meanwhile, the Babylonians in their fertile crescent-shaped valley between the Tigris and Euphrates rivers were making similar strides in mathematics. Less wrapped up than the Egyptians in concerns with life after death, they turned their knowledge of geometry to improving their present life. The Hanging Gardens of Babylon—also a wonder of the ancient world—could be enjoyed by King Nebuchadnezzar here and now, while the Egyptian pharaohs had to wait until they died to use the pyramids. The greatest structures in Babylon were not tombs but palaces and public buildings.

The Babylonians were great farmers and traders, and their financial dealings meant that they needed—and developed—a facility in computing numbers. As early as 4000 B.C. they had a basic arithmetic and soon were solving simple algebraic equations. By the time Babylonia was at its height, around 600 B.C., her mathematicians were attempting to prove theorems in geometry. They did not assume from direct observation, for instance, that the sum of the angles of a triangle is 180°. Instead they tried to give a logical proof. The Greeks are often given credit for being the first to construct proofs, but recent discoveries point to the Babylonians as the true originators. Whatever they did in offering proofs, however, has not survived and there is little evidence that they went very far in that direction.

Here the door of history closes on Egypt and Babylonia. Two great civilizations slowly fell into decay. The greatest was yet to come: Greece. Its Golden Age lasted but three centuries. Yet there have never been three centuries like them in the whole history of mankind. From a few Greek cities—actually small towns by today's standards—came the most spectacular intellectual achievements imaginable, achievements that laid the artistic, scientific, and political foundations for all of Western civilization. Such modern-sounding ideas as atomic physics, democracy, and communism all date back to ancient Greece. In her three golden centuries—from the beginning to the end—the entire citizen population of the Greek city-states was less than the present-day population of New York or London. In other words, a relative handful of people in a short span of years reshaped the whole path of human thought—and two thousand years later we are still on the same path.

The specialty of the Greeks was mathematics, and it is here that individual men begin to be identifiable on the mathematical scene. Pythagoras, one of the first, came from the island of Samos about 582 B.C. The exact date is unknown—indeed, most of what we know about him is conjecture, with a great many anecdotal fables thrown in.

Mathematics was hardly a science when Pythagoras began studying it. From Thales, another Greek mathematician and philosopher under whom he most likely studied, he learned about numbers, the building blocks of mathematics. These numbers did not include all the numbers modern mathematicians work with. Only the rational, positive integers were included in the system, that is, the numbers children first learn to count: 1, 2, 3, 4. . . . Negative numbers were not used, for reasons to be discussed later, nor were irrational and imaginary numbers. Zero was unknown and although fractions were used, they were not considered as numbers.

From Thales, Pythagoras also learned to operate on numbers—to add, subtract, multiply, and divide. Mathematics consists of operating on numbers in these different ways to get another number, which is usually called "the answer." "Thus number may be said to rule the whole world of quantity, and the four rules of arithmetic may be regarded as the complete equipment of the mathematician," said James Clerk Maxwell, a physicist of the nineteenth century.

When Pythagoras had learned as much as he could in Greece, he went to Egypt and Babylonia where he studied the collection of rules that passed as geometry. Then he returned home to find Samos in the hands of a tyrant, Polycrates. The mainland had been partly taken over by the Persians, and, feeling he would be better off elsewhere, Pythagoras settled in the Greek colony of Crotona in southern Italy. There he established a school more like a monastic order of mystics than a school as we know it. Only men were allowed to become members—although there may have been some nonmember women students; Pythagoras is reputed to have married one of them. The men were bound by strict rules. For the first five years they were not supposed to speak; at no time were they allowed to wear wool, eat meats or beans, use iron to stir a fire, touch a white rooster, or leave marks of ashes on a pot. If they failed to survive their low-protein diet, they could take solace in the belief that they would be reborn again, perhaps in another form. Pythagoras supposedly went walking one day and saw a man beating a dog. "Stop," he cried, "that's an old friend of mine you're whipping who has been reborn as a dog. I recognize his voice."

The basis of Pythagoras' philosophy was *number*. Number ruled the universe. Number was the basic description of everything. "Were it not for number and its nature, nothing that exists would be clear to anybody . . . You can observe the power of number exercising itself not only in the affairs of demons and gods, but in all the acts and thoughts of men," said one of his disciples.

One example of the power of numbers that the Pythagoreans discovered is that the pitch on a stringed musical instrument is dependent on the length of the string. A string twice as long as another will give a pitch twice as low.

Another less worthy discovery was the special character each number supposedly has. One, for instance, is reason; 2 is opinion; 4, justice. Four dots make a square and even today we speak of "a square deal" to mean justice. Odd numbers are masculine and even numbers feminine. Therefore, reasoned Pythagoras, even numbers represent evil; and odd, or masculine numbers, good. Five, the first sum of an odd and even number (1 was not counted) symbolizes marriage. On and on went this strange number lore whereby the future could be predicted, a person's character known, or any other vital bit of information revealed. Today we have our own unlucky 13 and our soothsayer's seventh son of a seventh son—all remnants of Pythagoras' ideas. While present-day numerologists may be in bad repute, they are in good company.

In spite of this superstition and mysticism that the Pythagoreans draped over mathematics, they made the first really great strides. Mathematics is usually broken down into four different fields—number theory or arithmetic, geometry, algebra, and analysis. Pythagoras did his work in the first two.

Number theory is that branch of mathematics that deals with the properties and relationships of numbers. Exotic, difficult, yet seemingly simple, it is the most worked-over and least developed field of all. Amateurs and professionals are on almost equal footing in this field which as yet has no foundation and very little structure.

Pythagoras was fascinated with number theory. Instead of worshiping the dead as did his Egyptian teachers, he worshiped numbers and tried to find out as much as possible about them. His fancy theories about numbers having special characters added little to the subject; his work with prime numbers—numbers which can be divided evenly only by themselves and 1, such as 2, 3, 5, 7, 11; perfect numbers—numbers equal to the sum of their divisors, such as 6 $(1 + 2 + 3 = 6)$; and amicable numbers such as 284 and 220, where the sum of the divisors of each number equals the other number $(1 + 2 + 4 + 71 + 142 = 220$ and $1 + 2 + 4 + 5 + 10 + 11 + 20 + 22 + 44 + 55 + 110 = 284)$, was really important. Only one pair of amicable numbers was known to the Greeks. The next pair, 17, 296 and 18,416, was not discovered until 1636. Today more than four hundred pairs are known, but the question of whether the number of amicable numbers is infinite still remains to be settled.

Prime numbers are as fascinating today as they were 2,500 years ago. Pythagoras, like his host of successors, tried to find a method of generating primes, or at least a test to determine whether a number is prime or not. On both counts he failed. Two and a half millennia later the problem is still unsolved, despite the fact that thousands of people have devoted millions of hours to it.

One interesting theorem about primes which was not set forth until centuries later is the Goldbach theorem, which states that every even number is the sum of two primes. No even number has been found that violates the theorem, yet no one has ever been able to prove it.

Primes are part of the scaffolding of the number system, but the structure of this scaffold is still obscure, still hidden despite the research of centuries. Prime numbers are imbedded throughout the system, yet no pattern in their apparently random appearance can be detected; no outward signs distinguish them from nonprimes; and no method exists to tell exactly how many primes occur in a certain number span. They are an infinite code that no one has ever been able to crack.

In geometry, "Pythagoras changed the study . . . into the form of a liberal education, for he examined its principles to the bottom and investigated its theorems in an immaterial and intellectual manner." In other words, he raised mathematics to the level of a science. He recognized the need for proving theorems. In a modern court case, proof must proceed from evidence; and so it was in geometry. To prove his theorems, Pythagoras had to start somewhere, with certain basic assumptions that themselves do not rest on other assumptions. Then, proceeding logically, he could prove the more complicated theorems by using only the basic assumptions and theorems that had already been proved with them.

The evidence needed in geometry, however, is different from the evidence needed in a court case, which rests on empirical facts or observation. The proof of one man's guilt in court does not mean that all men are guilty; but in geometry we must be able to prove that what is true of one triangle, for instance, is true of *all* triangles. This requires evidence of a different type. And what evidence is there? What assumptions can be accepted without having to be proved? Behind every assumption there are others and behind these even others. How far back can or should one go in establishing what are the basic assumptions? Eventually one reaches a point where the assumptions apparently cannot be proved; they do not seem to rest on others and must be accepted as "self-evident." It was these "self-evident" assumptions that Pythagoras used. Any man can see the truth of "the whole is greater than any of its parts" or "the whole is equal to the sum of its parts" without requiring a proof. Furthermore, these truths can be applied universally—they are always true. Every line is greater than its parts, every circle is greater than one of its arcs, and so on.

Thus Pythagoras made a deductive science out of a collection of rules that had been found by observation and trial and error. He put these rules on a foundation of logic from which mere observation was excluded. The importance of this contribution cannot be stressed too strongly, for Pythagoras' method of using simple, "self-evident" axioms to deduce more complicated theorems is basic to all science. He set geometry up as a model of formal logic —precise, neat, and perfect; as different from what it had been as cooking is from chemistry.

One of the most famous theorems arrived at through this deduc-

tive method and attributed to Pythagoras, the one that bears his
name, is that which gives the hypotenuse of a right triangle when
the lengths of the two arms are known: the hypotenuse equals
the square root of the sum of the squares of the other two sides.
Pythagoras' proof of the theorem is now lost—there are many
ways of proving it.

Geometry is number given shape. The Greeks may have had
some vague feeling that this was so, but they never saw it very
clearly. They managed to develop geometry so well without making the connection between it and numbers that for over fifteen
centuries men were blinded to the fact that geometry ultimately
boils down to numbers. This connection will be discussed more
fully in a later chapter, but for now it might be well to note that
a line can be said to be made up of units of any standard length,
each unit equal to 1. Increasing the length of a line is the same
as adding one number to another. Decreasing the length is comparable to subtraction of numbers, and so on.

The Greeks, however, were not interested in measuring their
lines; the study of geometry for them consisted of studying the
various forms—squares, triangles, circles, etc.—and the relationships between them and their parts. It is the idealized form, the
perfect square or triangle—not the particular—that was their subject matter.

This emphasis on form found its way into almost every aspect
of Greek life. The forms of tragic and comic literature were set
forth by Aristotle and adhered to even in Elizabethan times. In
sculpture, the idealized human form was the Greek goal: Venus
de Milo is not a particular woman, she is Woman, just as the
Greek triangles were not particular triangles but idealized, perfect ones. In architecture—indeed, in the whole plan of the
Acropolis—form was the guide. The proportions of the various
buildings were determined mathematically: "It may be said that
all parts of the Parthenon, the Acropolis, the city as rebuilt, the

walls, and the port, are expressible in the terms of a geometrically progressing series . . ." *

"God ever geometrizes," said Plato, and the Greeks dedicated themselves to discovering and extolling the geometry of the universe. Learning and knowledge, with their mathematical basis, excited in them all the passion that ordinarily is reserved for religion. Mathematics, to them, was perfection, a beautiful structure of logic whose truths have an independent existence completely apart from the existence of mere mortals. Even today, this Greek way of viewing mathematics persists among nonmathematicians. Ten and 10 are 20 and never anything else. We all know it and would stake our lives and fortunes on it. Yet hardly a week goes by that the average person does not add 10 to 10 on his watch or clock and get 8! Obviously, there are other ways of working with numbers than the usual one.

The Greeks' emphasis on geometry kept them from progressing in algebra and from perfecting a number notation system. They dealt with numbers geometrically whenever they could. To multiply 2 times 2, Pythagoras constructed a square with each side equal to 2 units. The area of the square is equal to the product of its sides, or 4. (We still call 4 "the square" of 2 and 8 "the cube" because these are the shapes that result in the Greek method of multiplication.) † Thus, geometry, arithmetic, and algebra can

---

* Robert W. Gardner, *The Parthenon, Its Science of Forms*, New York, 1925.
† The Greek predilection for converting all arithmetical processes into geometric ones resulted in some elegant and ingenious proofs. For instance, a quadratic equation such as $x^2 - xy - y^2 = 0$ was stated geometrically by Euclid: To cut a given straight line so that the rectangle contained by the whole and one of the segments shall be equal to the square on the remaining segment. Thus, if x is the given straight line, it can be cut into two segments, y and $x - y$ (Diagram 1). From the line x, or $x - y$ and y, one side of a rectangle can be formed. The other side, according to the stated problem, can be made from one of the segments, y or $x - y$. Let us use $x - y$ for the side of the rectangle (Diagram 1). From the remaining segment, y, a square can be made (Diagram 2). According to the problem,

deal with exactly the same problems stated in different terms. The
Greeks preferred to use the geometric method wherein numbers
were handled as forms. For this reason, not only their number
notation system but their field of numbers never developed.

Zero and negative numbers, for instance, are impossible in
Greek geometry, for a line cannot be zero or less in length, for
then it is no longer a line. The Greeks never even invented a symbol for zero. They had no use for it. Their number field included
only real, positive integers, and when they did discover another
kind of number, they retreated in horror.

This other kind of number is what is known today as an irrational number. Pythagoras discovered irrational numbers in working with his theorem for finding the length of a hypotenuse. All
applications of the thorem do not yield neat, whole numbers.
Some do not even yield fractions. Take the simple case of a right
triangle whose arms are each 1. One squared is 1. Therefore, the
hypotenuse is the square root of $1 + 1$, or 2; and the square root
of 2 is 1.414214. . . . Pythagoras used a fractional value rather
than a decimal one, as decimals were not invented until much
later, but he could nevertheless see that 2 has no square root—or

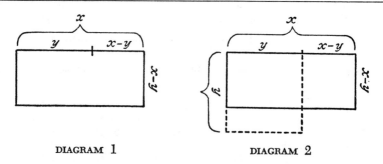

DIAGRAM 1   DIAGRAM 2

the square must have the same area as the rectangle, or, y^2 must equal x
times $x - y$. Setting this up as an equation, $x(x - y) = y^2$. Multiplying
x times $(x - y)$ we get $x^2 - xy$; transposing the y^2 to the left side of the
equals sign (which means that the sign of y^2 must be changed from plus
to minus), gives $x^2 - xy - y^2 = 0$, which is the algebraic statement of the
geometry problem.

at least not one that can be written down, for the number goes on to infinity. Although the hypotenuse of the right triangle has a definite magnitude or length, no number existed to describe it! Obviously, something was wrong. Pythagoras' whole wonderful religion of numbers threatened to come tumbling down all because of these numbers that are not really numbers. "Unspeakables," he called them and decided to suppress their existence. The members of his school were sworn to secrecy about this blight, this imperfection in mathematics. But the secret did get out and Proclos relates that "those who first divulged the secret of the unspeakables perished in shipwreck, to a man."

Today these unspeakables or irrational numbers are admitted to our number system, although many a mathematician is still uneasy about them. "God made the integers, the rest is the work of man," said Kronecker, a mathematician of the late nineteenth century. Like Pythagoras, he proposed banishing irrational numbers, believing them to be the source of mathematical confusion. A facetious cynic might say that irrationals are like women, we can't live with them but can't do without them. The digits of an irrational number are infinite, but does that mean that the number does not exist? It is not possible to write down the complete natural number system of integers either, for it also goes on to infinity. Yet who would toss out these rational numbers simply because they cannot all be written down?

The problem of irrationals is closely tied to that of infinity, and the knots of these problems are part of the warp and woof of the mathematical tapestry. They appear and reappear throughout mathematical history—as we shall see—and each time they crop up mathematicians react just as Pythagoras did. They ignore them or pretend they don't exist.

Pythagoras died around 507 B.C., supposedly burned to death in his school, which was set on fire by the local Crotonians who regarded the weird and exclusive Pythagoreans with suspicion. Some say that Pythagoras managed to escape, only to be murdered several years later in another town. At any rate, the fraternal order he had founded lasted two centuries before dying out, but the more reliable teachings never died. They became the foundation for other mathematicians to build on.

By the time Euclid was born in the late fourth century B.C., Greek geometry had come a long way from the state in which Pythagoras had left it. It had become more scientific and less mystical. Scores of new theorems had been added; curves, circles, and solids were studied as well as straight lines and plane figures.

The best-known school for mathematics and the closely related subject of philosophy was Plato's Academy.* Famous and respected, the Academy was the Oxford, Cambridge, Yale, and Harvard universities of its time all rolled into one. Here the golden flowering of Greece occurred, and here the greatest minds of antiquity assembled to lecture and listen. Established in 380 B.C., it survived repeated invasions, outlived tyrant after tyrant, and saw two great civilizations fall—the Greek and the Roman—before its doors were finally closed in the sixth century A.D. by the Emperor Justinian. Euclid is believed to have studied at the Academy, learning all that Greece had to teach about mathematics.

He was probably still at the Academy when Alexander the Great set out on his life's mission of conquering the world. Greece, along with Egypt and other Mediterranean and Hellenic states, fell to the Macedonian armies. Alexander put his various generals in charge of the vanquished countries, went off to new conquests, and, according to legend, sat down and cried when there were no more worlds to conquer. In 332 B.C., he established his capital at Alexandria in Egypt and nine years later died at the age of 33,

* "Let no man ignorant of mathematics enter here" is supposed to have been inscribed over the door of the Academy. This motto, however, may have been originated by a twelfth-century A.D. monk who, generously and without regard for fact, first attributed it to Plato. There are no previous records that connect the phrase with the Academy.

leaving his general Ptolemy to govern the capital. Ptolemy, a learned man, established not only a great dynasty that included Cleopatra, but also founded a great university that soon surpassed even Plato's Academy. He invited all the greatest scholars to come there to study and teach—among them was Euclid.

In Alexandria, Euclid found a flourishing, extremely modern and cosmopolitan city, more worldly and concerned with practical affairs than the slowly dying Greek cities. Although Euclid probably remained in Alexandria until he died, he never really adopted its ways. He was a Greek (although some say he was born a Phoenician) and thoroughly Greek in his outlook. The Alexandrians' practical application of mathematics to building pumps, fountains, and even steam-driven motors meant nothing to Euclid. He loved mathematics for its own sake, not for its practical uses.

There is a story that one of his pupils complained that learning theorems was pointless—they were of no practical value. Euclid commanded a slave to give the boy a coin so that he could make a profit in studying geometry.

On another occasion, so legend says, Ptolemy's son asked Euclid if there was not an easier way to learn geometry than by studying all the propositions. "There is no royal road to geometry," Euclid replied and sent the prince back to his books.

Outside of a few anecdotes and dubious dates, so little is known for certain about Euclid the man that it might be well to turn to Euclid the mathematician, where conjecture is replaced by fact.

With Euclid, Greek or Hellenic geometry rose to its highest. He represents the culmination of three centuries of mathematical thought, and his crowning achievement is the *Elements*, thirteen books in which he presents the basic principles of geometry and the statements and proofs of the theorems, systematized into a consistent whole. Euclid himself did not discover, or invent—depending on one's viewpoint—all the theorems presented in his *Elements*. He compiled the work of former mathematicians— Pythagoras, Eudoxus, Menaechmus, Hippocrates, etc., improving on the old proofs, giving new and simpler ones where needed, and selecting his axioms with care. Different schools of Greek mathematicians often used different axioms—some that were not really "self-evident." But Euclid selected only the most basic and obvious ones. He was careful to take nothing for granted. Everything was rigorously proved; and the result was a system of cold, logical, objective beauty. So well did he do his job that he was taken to task for being too petty. The Epicureans ridiculed him for proving so obvious a proposition as the one which states that the length of two sides of a triangle is greater than the third side. Modern mathematicians, however, charge Euclid with not being petty enough and with taking too much for granted. The flaws in his system did not come to light for many centuries and will be discussed in later chapters.

Almost immediately the *Elements* became—and remains—the standard text in geometry. When the printing press was invented, Euclid's *Elements* was among the first books to be printed. It is the all-time best seller as far as textbooks go, and wherever geometry is taught, there is Euclid.

Besides writing his definitive book on geometry, Euclid, like all Greek geometers, tried to solve the three classical problems of trisecting an angle, doubling a cube, and squaring a circle. For centuries the Greeks tried to solve these problems, using an unmarked ruler and a string compass.

The first problem seems simple enough and resolves itself into the cubic equation $4x^3 - 3x - a = 0$, where a is a given number. It can be done with certain marked instruments—but not with the unmarked ones the Greeks used. Although the Greeks did not know it, a ruler and compass can only be used to solve problems that resolve themselves into linear and quadratic equations, but not cubic ones. (A linear equation is one in which no term is raised to a power higher than 1; a quadratic equation contains terms raised to the second power, such as x^2, but no higher; and a cubic has terms raised to the third power, such as x^3.)

The second problem—doubling a cube—also resolves itself into a cubic equation, $2x^3 = y^3$, or even more simply, $x^3 = 2$. This problem, also insoluble with unmarked tools, has a pitfall into which many an amateur falls and which has been celebrated in legend. The Athenians, according to the story, once consulted the Delian oracle before undertaking a military campaign and were told that to insure victory they should double the size of the altar of Apollo, which was a cube. They duly built an altar twice as long, twice as wide, and twice as high as the original. Believing that they had fulfilled the oracle's request, they went confidently to war—and lost. Actually they had made the altar eight times as large, not two.

The third problem, squaring the circle, is not possible at all with any kind of instrument, nor can it be stated algebraically. The problem involves finding an exact value for π (pi), the ratio between a circle and its diameter. But π is an irrational, transcendental number for which there is no exact value, and anyone who tries to find an exact value is simply wasting his time, for it has been proved that it cannot be done.

Although the Greeks failed in their attempts to solve these three problems, they were led to explore different areas of geometry, mainly conic sections, in their search for solutions.

Euclid also attacked the old problem of irrationals that had given Pythagoras so much trouble and even succeeded in proving their existence. Taking a right triangle whose arms are each 1, he showed that there is no rational solution for the length of the hypotenuse, which is $x^2 = 2$. Euclid first assumed that a solution could be found. The solution would have to be a number in its lowest terms, p/q, such that $p^2/q^2 = 2$. Either p or q or both must be odd, for if they are both even, the fraction can be divided by 2 and is not in its lowest terms. Euclid then showed that p cannot be odd by substituting p/q for x in the original equation, $x^2 = 2$, which gives $p^2/q^2 = 2$. Transposing, we get $p^2 = 2q^2$. Since p^2 equals twice another number, p^2 must be even; and since the square root of an even number is also an even number, p must be even.

Therefore, if a solution is possible, q must be odd. We already know that p is even and therefore is equivalent to twice another number, or $2r$. Substituting $2r$ for p in the previous equation $p^2 = 2q^2$, we get $4r^2 = 2q^2$, or $2r^2 = q^2$. Therefore q, for the same reasons as stated for p, is even. Yet the fraction p/q was assumed to be in its lowest terms. But if both p and q are even, the fraction cannot be in its lowest terms. Thus, nothing but absurd contradictions result from the assumption that the problem can be solved rationally, i.e., with the numbers the Greeks used. No ratio exists between the numbers p and q such that $p^2/q^2 = 2$. For this reason the square root of 2 is called irrational. Irrational simply means that the number has no ratio, not that it is insane.

In number theory, Euclid, like most mathematicians of his day, studied primes, searching for a test to determine whether any given number is a prime or not. Needless to say, he never found it, but he did settle one question about primes: whether or not they are infinite in number.

Any half-observant schoolchild can see that primes, like the atmosphere, tend to thin out the higher we go. From 2 to 50 there are fifteen prime numbers; from 50 to 100, only ten. It seems entirely possible that primes might thin out and eventually disappear completely. Euclid devised an ingenious proof to show that this is not so; the number of primes is infinite.

If the largest prime is n, he reasoned, then there must be another number larger than n which can be generated by multiplying 1 times 2 times 3 and so on up to n and then adding 1 to the result. In mathematical symbols this number would be written $n! + 1$. (Read aloud, "n factorial plus one.") Now if n is the largest prime, this new number is not a prime. If it is not a prime, it must have a divisor smaller than itself. But if it is divided by any number up to and including n, there will always be a remainder of 1, since the number was formed by multiplying all the numbers from 1 to n together and then adding 1. Therefore, its divisor—if it has one—must be greater than n and that divisor itself must be a prime, which is a prime number greater than n. If $n! + 1$ has no divisors greater than n, then $n! + 1$ itself is a prime greater than n. In either case, a larger prime than n exists. Therefore, the number of primes is infinite, for in the same way a larger prime than $n! + 1$ can be found, and so on ad infinitum.

This proof of the infinity of primes—which also presupposes an infinity of numbers—and the proof of the existence of irrationals did not spur Euclid and his colleagues on to a fuller investigation of the nature of infinity. On the contrary, they avoided working with infinity and irrationals—which are part and parcel of infinity, for an irrational number is itself infinite. Pythagoras had become so upset by irrationals that he tried to suppress them. Euclid, while not as extreme in his reaction, ignored the whole idea of infinity and infinite irrationals, preferring to deal with finite, static shapes that stood still and did not go bounding off into distant space.

The problem of infinity, however, is not one that can be ignored. No matter how much it is written out of the script and shunted off into the wings, it constantly creeps back on stage to shout, "Here I am! What part do I play in mathematics?" For 2,500 years no one knew, so they pretended they did not hear.

Euclid died, no one knows exactly when or where, but the famous school at Alexandria continued. One of its greatest—if not the greatest—students was Archimedes, a Greek boy from Syracuse who attended the school during or shortly after Euclid taught there.

Born about 287 B.C., Archimedes is popularly known as "the Father of Mathematics" and ranks with Newton and Gauss as one of the three greatest mathematicians who ever lived. Unlike his two peers, his background and upbringing steered him directly toward mathematics. His father, Pheidias, was a well-known

astronomer and mathematician. Since Archimedes belonged to
the leisured upper class—he may even have been related to King
Hiero II of Syracuse—a thorough grounding in mathematics was
necessarily part of his education. It was in the economic and social
order of things that he become a mathematician of some kind, if
only a bumbling amateur. But it was genius—genius of a sort
that has cropped up only three times in recorded history—that
made Archimedes the mathematician he was.

After studying at Alexandria, Archimedes returned home to
Syracuse, where he spent the rest of his life. He continued to
correspond with three mathematicians at Alexandria, Conon,
Dositheus, and Eratosthenes, with whom he discussed various
problems and their solutions.

At Alexandria, Archimedes had developed a taste for applied
mathematics, although he preferred to devote himself to pure
theory. The men of the Hellenistic world to which Archimedes
belonged were more worldly and materialistic in their outlook
than the men of the pure Greek or Hellenic world. Practical problems were not beneath them, and in applied mathematics Archimedes made a distinguished and celebrated reputation.

No problem seemed too small for him. When King Hiero asked
him to find out whether a golden crown made by a none too reliable goldsmith had been adulterated with silver, Archimedes devoted as much thought to it as to more important problems. One
day while sitting in the bathtub, he suddenly hit on a solution.
In his excitement over his discovery, he sprang from the tub,
dashed outdoors, and ran through the streets shouting, "Eureka,
eureka!"—"I have found it, I have found it!" To the astonished
Syracusans it probably seemed that the naked man had lost something—either his clothes or his mind.

What Archimedes had found was the first law of hydrostatics:
that a floating body loses in weight an amount equal to that of
the liquid displaced. Since equal masses of gold and silver differ
in weight and therefore displace different amounts of water, it

was a simple matter for Archimedes to determine whether the crown was pure gold or not. He simply had to weigh the crown—immerse it in water and weigh the water displaced. The ratio of the two weights should be the same as the ratio between the weights of a piece of pure gold and the water it displaces. Unfortunately for the goldsmith, the ratios were different, indicating that the crown was not pure gold.

Apocryphal or not, the tale does shed some light on Archimedes' personality. He was a man capable of intense concentration and when deep in thought was oblivious to everything around him. Mathematics was his passion, his all-encompassing interest, and he used every moment to study it. He would sit by the fire and draw diagrams in the ashes. After his bath, when he oiled his skin as was the custom, he would become absorbed in working out problems, using his oiled skin as a slate and his fingernail as a stylus. When concentrating on a problem he thought of nothing else, forgetting to dress himself, to eat or sleep, and, according to Plutarch, had to be "carried by absolute violence" to the table or tub.

In theoretical mathematics, Archimedes concerned himself with measuring areas and segments of plane and solid conic sections. In this work he made his greatest contributions. He cleared up almost every geometric measuring problem left. When he was through, mathematicians, like Alexander the Great, had to look for new worlds to conquer.

His proof that the volume of a sphere inscribed in a cylinder is two-thirds the volume of the cylinder he considered to be one of his greatest discoveries and requested that his tombstone be engraved with a sphere inside a cylinder and the ratio 2 to 3.

He discovered formulas for finding the areas of ellipses and segments of parabolas, and he attacked the problem of finding the area of a circle, which is the old problem of squaring a circle in another guise.

Archimedes' approach to the problem of measuring a circle—

which involves finding π—is extremely important, for it contains
the central idea of infinitesimal analysis. He drew a circle and
then inscribed and circumscribed hexagons (Diagram 3). Obviously, the area of the circle lies somewhere between the areas
of the two hexagons. By doubling their number of sides, Archi-

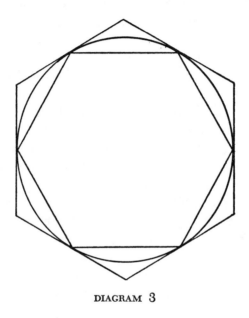

DIAGRAM 3

medes could bring the perimeters of the polygons closer and
closer to the circumference of the circle. At first the polygons
touched the circle at 6 points each, then 12, then 24, and so on.
By continuing the process, every point on the circle would eventually correspond to every point on the polygons. How many points
are there on a circle? An infinite number. Archimedes would have
had to continue to double his polygons an infinite number of times
to determine the exact area of the circle. He stopped when he
had polygons of 96 sides, determining that π lies between 3⅐ and
3¹⁰⁄₇₁, which is correct to two decimal places. Later mathematicians, using the same method, computed π to as many as thirty-

five decimal places. This was done in the seventeenth century by Ludolph van Ceulen of Germany, who used polygons having 2^{62} sides. He worked almost to his dying day on the problem and was rewarded by having the number named after himself and inscribed on his tombstone. The "Ludolphian number" in Germany is our π.

A later mathematician by the name of William Shanks persevered to the 707th decimal place, laboring two decades over the problem. He had the misfortune, however, to make an error on the 528th decimal, thus rendering the last few years' work and couple hundred decimals invalid.

Today electronic computers can give π to thousands of decimal places in a few minutes. But no matter to how many thousands or even millions of places π is calculated, there is no end.

In using the method of infinite processes to determine areas of parabolic segments, Archimedes was more successful than with the circle, for his limiting number was not the irrational π but a rational number that could be "caught" exactly between the converging polygons. Further, he invented a method of determining the limit of a converging geometric series such as $1 + \frac{1}{2} + \frac{1}{4} + \frac{1}{8} + \frac{1}{16} + \cdots$.

Archimedes' method of calculating π is important not only because it gives a fairly accurate result, but because it embodies the elements of calculus: infinite processes and limits. One more step and he would have been inside the door that was opened almost two thousand years later by Newton.

There were ancient mathematicians who believed that π or any irrational number could be determined exactly by infinite processes. That is, they did not believe that infinity is really infinite, but that it could be reached with hard work and supreme patience. Infinity was the uncounted, not the uncountable. Archimedes was not among these optimists. He emphasized very graphically that infinity is beyond our reach. In a short book called the *Sand Reckoner*, he wrote: "There are some who think that the number of the sands is infinite in multitude . . . And again, there are some who, without regarding it as infinite, yet think that no number has been named which is great enough to exceed its multitude." He then proceeded to calculate the number of grains of sand needed to fill the universe, using a number system of his own invention, and noted that even this number, 10^{63}, is short of infinity, which can never be reached.

In the "sordid" and "ignoble" field of applied mathematics, Archimedes' work in hydrostatics has already been mentioned. Further requests from the king resulted in the invention of pulleys and levers that could move tremendous weights. After demonstrating himself that his inventions, operated by one man, could launch a fully loaded ship, Archimedes declared, "Give me a place to stand on and I can move the earth."

Archimedes' inventions were put to good use, for when the Roman general, Marcellus, tried to invade Syracuse in 212 B.C., there was the most astonishing array of machines waiting for him: cranes and lifts that reached out and hoisted ships from the sea, missile launchers that could toss 500-pound stones at the invaders. Archimedes even harnessed the energy of the sun to set fire to the enemy's ships.

The Romans retreated in terror, hardly knowing what had happened. Urged on by their officers, they attacked the walls again and were met with a barrage of stones and arrows, and "became so filled with fear that if they saw a little piece of rope or wood projecting over the wall, they cried, 'There it is, Archimedes is training some engine upon us,' and fled." But before long they were back, and instead of attempting an invasion from the waterfront, attacked the rear of the city via land. The Syracusans, who were celebrating a festival that day, fell to the Romans with almost no resistance.

Archimedes, as usual, was absorbed in working out a geometry problem in the sand. A Roman soldier approached him and stood near by, casting his shadow on the diagram. The old mathematician—he was seventy-five—asked the soldier to move out of the way, and the angry Roman ran him through with a sword. Another version of Archimedes' death says that the soldier ordered him to go to Marcellus' headquarters. Archimedes told the soldier to wait a minute—he wanted to finish his problem first—and was then run through.

With the death of Archimedes, mathematical progress came to a virtual standstill. He had developed the subject to the point where no further advances could be made without algebra and analytic geometry. History had to wait seventeen centuries for the next major step.

The Greeks had started with an almost barren world and formed from it geometry. They had laid the foundations and set the standards for all of science. They had completely mastered geometry and made a start in trigonometry. Archimedes pointed the way to calculus or analysis, but there was no one to follow.

The Roman Empire was beginning to flower—indeed, it was growing like a weed, choking out the Hellenistic cities and with them the Greek thirst for knowledge. When Caesar invaded Egypt, he set fire to the ships in the harbor of Alexandria. The fire got out of control, spread to the library, and burned half a million manuscripts—the repository of all ancient knowledge. What the flames missed, the Moslems looted seven centuries later, scattering the heritage of Greece to the four winds.

Meanwhile, the world saw the short blazing glory of the Roman Empire before what has been called "the midnight of history" set in with the Dark Ages.

Cardano
1501-1576

The Renaissance burst upon Europe in Italy and then spread
to the rest of the continent, scattering the cobwebs of the medieval period, breathing new life into art, science, and commerce. Vigorous, brilliant, ambitious, worldly men replaced the
illiterate serfs of the Middle Ages whose main concern had been
the salvation of their souls. With the Renaissance, men's eyes
turned earthward, and here, rather than in heaven, men sought
their immortality. Explorers risked their lives for gold and a foun-
tain of youth; alchemists brewed "elixirs of life," tried to turn
base metals into gold; and the rich sent their graven and painted
images down into unknown posterity.

Girolamo Cardano belonged to that age—a supreme product of
it. His eagerness for earthly immortality became the wellspring
of all his actions. Like the ancient Greek who had burned the
temple of Diana to gain fame, Cardano did not care what posterity thought of him—just as long as it *thought* of him. And during
his lifetime he had good reason to believe that his ambition would
be fulfilled, for he was one of Europe's most famous and fashionable physicians, a noted mathematician and astrologer (the latter
profession had a recognized standing in the sixteenth century),
and the most widely read scientific writer of his day. He doctored
kings and cardinals, cast fortunes for princes and popes, pub-

lished original work in mathematics, and altogether wrote more than four hundred books, of which almost one hundred and fifty were printed. Today he is a mere footnote in the histories of medicine and mathematics and almost nothing he wrote is read.

Yet he typifies to the extreme the early Renaissance mathematician, whose main contributions to the science were in resurrecting it rather than in adding to it. For this reason alone, Cardano deserves attention.

In Europe very little of importance to mathematics had occurred since the days of Archimedes. Alexandria had had one last burst of splendor during which time her mathematicians measured the earth's diameter and the distance from the Earth to the Moon. Then scholarship was stifled by the practical-minded Roman Empire which flourished for another five centuries but added almost nothing to mathematics. When Rome fell, her ruin was as great as her glory had been. Gibbon in his *Decline and Fall of the Roman Empire* tells of how two servants during the reign of Pope Eugenius climbed the Capitoline Hill in Rome and described what they saw:

> The sacred ground is disfigured with thorns and brambles . . . The path of victory is obliterated by vines, and the benches of the senators are concealed by a dunghill . . . The forum of the Roman people . . . is now enclosed for the cultivation of pot-herbs or thrown open for the reception of swine and buffaloes. The public and private edifices that were founded for eternity lie prostrate, naked and broken, like the limbs of a mighty giant.

That was the state of Rome, and the rest of Europe was Rome writ large. Only in the East—in Turkey and India—was there intellectual light. Constantinople preserved Greek learning but added little to it. In India, however, three of the greatest strides in the development of mathematics were being made.

Our present number system was created, wherein only ten

digits, including zero, could easily and clearly express any number desired. It is impossible to estimate the importance of this
invention. Without it, mathematics could have progressed no
further.

The second development was the extension of the number field
to include negative numbers.

The third was the invention of algebra. Symbols were invented
to indicate the different operations—addition, subtraction, etc.—
and simple linear and quadratic equations in one unknown were
solved, using the same symbol for zero as for the unknown.

During the ninth and tenth centuries, the mathematics of India
spread to Arabia, where it was developed even further. One of
the most famous Arabian mathematicians was Omar Khayyám,
known to the Western world as the author of *The Rubaiyat.* Omar
Khayyám not only absorbed all that India had to teach but added
something of his own. He began the use of graphs to combine
algebra and geometry and may even have worked with the binomial theorem, thus anticipating Descartes and Newton by
several centuries.

Meanwhile, Europe was still in the Dark Ages. Not until the
Crusades did the knowledge of the Arabs begin to seep into Europe. Algebra and Arabic numbers were introduced to Italy in
1202 by Leonardo Fibonacci, an eminent Pisan scholar. Everywhere, the resistance against using the new number system was
great. In some places Arabic numbers were actually banned by
law. Slowly, however, Roman numerals gave way to Arabic ones,
except in places where rapid calculating does not matter, such as
on monuments, buildings, and watch faces, where they persist
even today.

It was not until the fourteenth and fifteenth centuries that the
Renaissance butterfly, brilliantly colored and alive, really began
to emerge from the dull, dark cocoon in which she had been
sleeping for almost a thousand years. Arab learning took root
and began to be taught in the new universities. Greek mathe-

maticians began migrating to Europe from Constantinople, driven by poverty and the invading Turks, who finally captured the city in 1453. With the rise of commerce and the need for bookkeeping and accounting, interest in numbers grew.

The year is now 1501, and the Renaissance is underway. In a little town not far from the bustling city of Milan in northern Italy, Girolamo Cardano was born "half-dead" on September 24. "I was revived in a bath of warm wine," he says, and also notes that because of the position of the stars and planets, "I ought to have been born a monster." Instead, he escaped only slightly maimed.

His father, Fazio, was an educated man whose round shoulders were attributed to hunching over his books too long. Although a lawyer by profession, he gained prominence as a "gifted and exceptionally skillful" mathematician. Leonardo da Vinci was known to have consulted him, probably about problems in perspective, which is related to geometry. Despite his learning, Fazio fared rather poorly in the scramble for wealth that characterized northern Italy at the time. He lived only sporadically with his wife, supposedly because he was not able to support her, although neighbors suspected that the real reason was because they were not married.

From his father, Cardano received his first lessons in reading and mathematics, as well as an introduction to the law. As a young boy he became his father's page, with the obvious conclusion that a legal career was planned for him. But the strain was too much for Girolamo. He became ill and was excused from work until he was thirteen.

On the whole, Cardano could boast of a childhood filled with unhappiness and bad luck. He was constantly sick, and his hot-tempered parents whipped him for the slightest disobedience. When he was seven they gave up beating him—whether because the whippings had done their job, or because they were useless, Cardano does not say.

His first conscious craving for fame came while he was still a teenager. One of his best friends died suddenly and Cardano was left with a deep impression of the transitoriness of life. What could one leave behind but a name or heirs? His friend had no heirs, and in a matter of months his name was no longer mentioned.

Cardano cast about for some insurance that the same thing should not happen to him. His parents had endowed him with nothing but "misery and scorn"; he had no money and little learning—the fact that he did not know Latin seemed especially galling; physically, he was weak and puny. All this was hardly the basis for notoriety, let alone fame. On the credit side, he had a good mind and "an unshakable ambition." He plainly states in his autobiography that "I am considered an opportunist," and the best opportunities seemed to lie in becoming a doctor. His father wanted him to become a lawyer, but after thinking it over, Cardano decided that law had done little for his father in the way of worldly esteem or gain. Medicine seemed to be more lucrative and was of "a sincerer character"—although the latter was hardly a quality to sway the ambitious Girolamo. Another reason he was drawn more to medicine than to law is that medicine probably appealed to his hypochondriacal nature. And so, Cardano decided that if he were to make his mark in the world, it would have to be as a doctor.

His decision evoked the usual family quarrels, with his mother, Chiara Michina, siding with her son. She had lost her first three children from a plague and wanted nothing but the best for dear Girolamo. By nature "fat, pious and quick-tempered," she raged and argued until her husband gave in.

At nineteen Cardano entered the local academy in Pavia and when it closed shortly afterward because of civil strife, he enrolled in the University of Padua. It was there that he won his first laurels. He was elected Rector, a purely honorary position, and made a name for himself as a debater. He modestly claimed that he defeated all his opponents not because of any special oratorical talent but because of "a certain superiority and perfection in reasoning." Then, lest history not give him his due, he added that "I was so exceptionally clever that everyone marvelled at my extraordinary skill," and no one wanted to debate against him.

After receiving his degree, the brilliant but insufferably conceited young doctor returned to Milan and applied for a license to practice medicine. The College of Physicians turned him down on the grounds that he was of illegitimate birth. Cardano protested that this was not so, that his parents were married but had not always lived together for financial reasons. Nevertheless, he was denied a license and, rebuffed, settled in the little town of Sacco, near Padua. Here he spent what he called the happiest years of his life. He married a local girl, Lucia Bandarini, and had three children by her, Giambatista, Chiara, and Aldo—three policies for immortality.

Cardano may have supported his family in part by lecturing and practicing medicine on the sly, but most of his income came from gambling—a habit he had picked up at the university. Not a day went by that did not see him tossing dice or shuffling cards. He knew every legitimate and illegitimate trick, although he claimed never to have used the latter. Good luck and a knowledge of the laws of probability, he said, accounted for his constant winnings.

Although in later years Cardano excused his gambling by saying that he had been driven to it because of financial necessity, he gambled almost every day for the rest of his life—long after the financial strain had disappeared. It was true, however, that during his early years he had so little money that at one point he was reduced to accepting handouts from the local poorhouse.

While at Sacco, Cardano was invited to lecture in medicine at the University of Pavia, which had again opened its doors. He turned the offer down because "there was almost no chance of getting paid."

His third application to the College of Physicians in Milan was accepted, most likely through the good offices of a prominent Milanese whose child Cardano had saved from death.

Cardano was now thirty-eight years old, and his career was just beginning. He threw himself into his medical practice and within two years was one of the most successful doctors in Milan. He lectured at the university, resumed his study of mathematics, began tidying his notes for several books, and acquired a patron who later became the Archbishop of Milan.

Now that his fortune had taken a turn for the better, Cardano liked to ruminate on the idea of how close he had come to dying before he had had a chance really to live. While still in school he had taken a trip on Lake Garda when a storm suddenly came up. The swamped and sinking boat miraculously made it to shore where Cardano ate a hearty supper while the rest of the passengers were still too excited and upset to eat at all. In his autobiography, he dwells on a good number of other close calls, leaving the inference that his survival was due to a guardian angel who he believed protected and warned him of imminent disaster. He notes that several times he crossed the street for no purpose, only to have a large chunk of masonry come crashing down in the exact spot he had just been standing. Only some sort of supernatural power, he was sure, could have saved him so often.

And why should he—along with several other great men, including Socrates—he blessed with a guardian angel? Among the possible reasons he dredges up are his "infinite love of truth and wisdom," his "great desire for justice," or, most incredible of all, the fact that "I give God credit for my success and never take any myself."

Whether Cardano or his guardian angel was responsible for his success really does not matter. But that a supposedly educated man should be so superstitious is almost incredible, especially a man who believed in and furthered the practice of the scientific method. Granted, his was an age of superstition—an age when every royal court had its astrologer. But Cardano believed all sorts of nonsense that even an astrologer would have laughed at. His parents had been excessively superstitious; Cardano himself admits that their ideas and old wives' tales were wild and improbable. Only later, when he grew up, did he begin to collect his own baggage of omens, never realizing that since he was looking for omens, he would find them.

He claimed that every important event in his life was presaged by strange dreams, howling dogs, sparks of fire, crowing roosters, cawing ravens, grunts, knocks, bangings, buzzings, or thumps. For instance, one night he had a dream about a strange girl he had never seen. A week later, he claims, he saw the same girl in the street and a few months later she became his wife.

The day his first son was to be baptized, a wasp flew into the room where the mother and child lay, circled the baby and then became trapped in the draperies. Years later Cardano was to remember this incident and construe its terrible import.

He also maintained that when people talked about him his ears buzzed. If something good was being said, his right ear buzzed; if something bad, his left.

He claimed that his knowledge came to him in "intuitive flashes." While at school he had bought a Latin book because of its lovely gold decorations, although he could not read a word of it. "The following day," he says, "I was as well versed in Latin as I am now."

The list of Cardano's naive and superstitious ideas could go on and on. Many a great man has had a soft spot in his head or heart for a particular bit of pseudo-scientific nonsense, but this does not and should not negate his real contributions. And Cardano's contributions were worthy enough. In mathematics alone, he wrote twenty-one books, eight of which were published. His *Ars*

Magna, the best of these, was the most complete treatment of algebra in the middle of the sixteenth century.

Geometry had not developed beyond the state in which the late Alexandrians had left it when Cardano began his studies. And it would be another century before any significant additions would be made. Algebra, however, was beginning to outgrow the swaddling clothes in which the Arabs had wrapped it; and Cardano was one of the men who helped it grow. This infant algebra was still too weak to stand, its skeleton barely defined, and its clothes ingeniously sewn together from a patched-up bag of tricks. In other words, algebra had none of the precision nor the beauty of Greek geometry. It was simply an assortment of rules and methods that could be used to solve mathematical problems and puzzles.

Whereas geometry had been food for sublime minds unconcerned with worldly affairs, algebra was the bread and butter of the marketplace. It had developed, in part, from accounting shorthand and grown up without a plan or structure, without axioms and postulates. Yet it worked, and men kept adding new things to it without even feeling the need of a logical foundation. Geometry might be compared to the well-brought-up, disciplined child whose formal manners delight. Algebra, on the other hand, grew up in a household where the parents were too busy making a living to administer discipline or teach manners. Not until the nineteenth century, did anyone notice or care that algebra lacked the polish and style of the more aristocratic geometry.

Cardano, as mentioned before, helped algebra grow. Of his twenty-one books on mathematics most were compendia of all that was known about geometry, algebra, and number theory. His *Ars Magna,* the most complete of these, was one of the great mathematical books of its day. He had already begun work on it when a new discovery came to his attention—a rule for solving cubic equations. Until the early sixteenth century only linear and quadratic equations could be solved algebraically, that is, equations of the first and second degree. Equations in which cubes occurred, such as $x^3 + x^2 - 4 = 0$, could not be solved. The difference between these three types of equations, linear, quadratic, and cubic, is tantamount to the difference between a line, a plane, and a solid in geometry. And just as solid geometry is generally more complex than plane, so cubic equations are more complex than linear or quadratic ones.

About the year 1505, Scipio Ferreus of Bologna found a rule for solving one type of cubic equation. He passed the rule on to one man—in those days the practice was not to make discoveries public but rather to keep them secret for private gain. Mathematicians challenged each other to solve problems, often with a cash prize posted by the competitors or an interested third party. These contests were popular even after Newton's day and were a source of great prestige as well as extra income for the winner.

In 1535, Ferreus' pupil challenged a man known as Nicolo Tartaglia, "the stutterer," to a contest that involved solving one type of cubic equation. Tartaglia, a self-taught scholar, accepted the challenge and managed to discover the rule for solving *any* cubic equation. Being the only man in the world armed with this knowledge, he could of course defeat any opponent.

Four years later Cardano graciously asked for the rule so that he could include it in his book, *Ars Magna*. Tartaglia replied, "When I propose to publish my invention I will publish it in a work of my own." This is the answer Cardano should have expected, but he became furious and wrote back calling Tartaglia a "great ignoramus, presumptuous and conceited." What he had hoped to achieve by flattery could hardly succeed by insult, and Tartaglia answered by again refusing to divulge the secret but challenging Cardano to a contest. This time Cardano answered in a politer tone, saying that he had written the nasty letter "hoping to cause what has happened to happen—that is, to have your answer together with the friendship of a man so singularly able." He neatly sidestepped the issue of a contest in which he knew he would be defeated and dropped a few hints about his new patron, the Marquis dal Vasto, who was interested in and eager to meet Tartaglia. Again he asked for the solution to cubic equations and promised to keep it secret.

Apparently the Marquis was the bait needed, for Tartaglia came to Milan—unfortunately on a day when the Marquis was out of town—and Cardano had his chance to use his charms. After swearing "by the sacred Gospel, on the faith of a gentleman" and "as a true Christian" never to divulge the rule, he was given the secret. Tartaglia returned home without ever seeing the Marquis.

Six years later the *Ars Magna* was published—and there for all the world to see was the solution Cardano had promised never to reveal. In his biography, written many years later, he had the effrontery to state that "I have never . . . divulged the secrets of my former friends." The one good thing that can be said for him in this affair is that in his book he generously credited Tartaglia with discovering the rule. Ironically enough, however, history has honored Cardano by naming the rule "Cardano's formula." With the publication of the rule, Cardano opened up a whole new category of equations for mathematical study.

Although he cannot really be credited with finding the solution to cubic equations, he can be cited for equally worthy advancements. He gave the first clear picture of negative numbers, first recognized the existence of negative roots, and made the first note of imaginary numbers.

His achievements here, and how he made them, can be clarified by a simple discussion of algebraic equations. An equation is a statement of equality, that is, the terms to the left of the equals sign equal the terms to the right. Using only one unknown, such as x, an infinite number of equations can be formed. Yet in every case, all these equations have one general form, known as the general equation:

$$ax^n + bx^{n-1} + cx^{n-2} + \cdots + px + q = 0.$$

The letters a, b, c, . . . in this general equation stand for given numbers or coefficients. The exponents n, $n - 1$, . . . stand for the highest power that x is raised to, the next highest, and so on until there is a term q, in which x is raised to the zero power (any number raised to the zero power is equal to 1). If there are terms missing, such as the x^2 term in $3x^3 - x + 10 = 0$, the equation still fits the general equation form, for the coefficient of x^2, the missing term, can be said to be zero. The x^2 term can be inserted by making the equation read $3x^3 + 0x^2 - x + 10 = 0$. Thus, any linear, quadratic, cubic, quartic, quintic, or higher-degree equation fits one general equation form.

Single equations in more than one unknown, such as $x - y = 30$, are known as Diophantine equations and have an infinite number of solutions. Values of 34 and 4, 35 and 5, 40 and 10, and so on can be taken for x and y respectively. Generally, in order to solve equations in 2 and 3 or n unknowns, the same number of equations must be given as there are unknowns. That is, to solve for two unknowns, two equations must be given; to solve for three, three equations; and so on. These are called systems of equations and are special cases of the general equation, to which they must be reduced by a simple process before they can be solved.

To return to the equation in one unknown, which is the main type handled at Cardano's time, only linear and quadratic equations were solvable. Then, through the reluctant generosity of Tartaglia, Cardano showed that cubic equations could also be solved; and shortly after, a protégé of Cardano's discovered solutions for quartic equations.

In solving even simple linear equations, Cardano and his contemporaries discovered that the limited number field passed on to them by the Greeks was not sufficient. Not every equation has a positive integer or even a fraction as a root, that is, as an answer. An equation such as $x + 10 = 0$ cannot be solved with positive integers or fractions, for x equals -10. Early mathematicians did not consider negative numbers to be numbers and any equation having a negative root was considered insoluble. Thus, a restriction was placed on negative numbers, just as an earlier restriction had been placed by the Greeks on geometric constructions. Gradually, however, men began to accept the idea of minus quantities as being real, and with this restriction removed, a great many more problems could be solved. Yet even with the recognition of negative numbers, only the positive roots of quadratic equations were found. For instance, the equation $x^2 - 4 = 0$ has two roots, 2 and -2. Only the former was noted by Renaissance mathematicians. It was Cardano who pointed out that the negative root must also be considered.

The way in which imaginary numbers and imaginary roots were discovered parallels that of negative numbers and roots. At first imaginaries, too, were considered ridiculous and impossible absurdities. Yet they kept cropping up as solutions. The Hindus and the Arabs rejected them in much the same way that Pythagoras had rejected irrationals. Yet irrationals had gradually made their way into the number system and negative numbers, too, had later been admitted. Cardano took the first step toward the admission of image numbers in his book *Ars Magna*, when he discussed the set of equations $x + y = 10$, $xy = 40$, which is the same thing as $x^2 - 10x + 40 = 0$.* He showed that the roots would be $5 + \sqrt{-15}$ and $5 - \sqrt{-15}$. Although he dismissed these expressions as being meaningless, he did give them some sort of substance merely by stating them. And as had happened with irrationals and negatives, the meaningless eventually gave way to substance.

In addition to his epochal *Ars Magna*, Cardano wrote a book on gambling that included the first exposition of the laws of probability. Before him, no one had ever attempted an investiga-

---

* The two equations are made into one by multiplying all the terms of the first one by x, which yields $x^2 + xy = 10x$. The second equation, $xy = 40$, is then subtracted from this equation and the terms of the remainder transposed.

tion of the subject—or even suspected that chance is not really chancy. But Cardano, driven by an inquisitive mind and acquisitive motives, began to realize that certain laws govern the toss of die or turn of cards and that one's winnings can be augmented by more honorable means than cheating. Like all real gamblers, however, he often ignored his own advice and trusted to luck, or, as has been charged, marked decks and loaded dice.

Cardano's book on probability fell into a literary limbo shortly after its publication—perhaps because its potential audience was largely illiterate. No further development of the subject was made until a hundred years later when Pascal and Fermat began their own investigations. Ironically, they are credited with founding this branch of mathematics while the true discoverer gets credit only for what he stole from Tartaglia!

As Cardano's medical practice prospered, he became even more prolific in his production of books, many of which were based on notes he had busily gathered as a student and gambler. He wrote on medicine, astrology, physics, psychology and mental health, among other subjects. The sheer bulk of his writings alone would have been enough to keep a healthy, energetic, ambitious man busy from morning to night for a lifetime. And Cardano was far from healthy—or so he would have us believe. He was a walking casebook of real or imagined ills, being plagued with catarrh, indigestion, congenital heart palpitations, hemorrhoids, gout, rupture, bladder trouble, insomnia, plague, carbuncles, tertian fever, colic, and poor circulation, plus various other ailments. He noted that "even in the best of health I suffer from a cough and hoarseness." Quite literally, he enjoyed poor health and seemed to take a neurotic pride in his list of infirmities, although he constantly grumbles about them throughout his biography. Yet his hypochondria (an ailment omitted from his list) shows through when he says, "It was my custom when I had no other malady to turn to my gout as a complaint." He admits a masochistic delight in "biting my lips, twisting my fingers, or pinching my arm until the tears started in my eyes"; but nevertheless pampered himself by spending at least ten hours a night in bed.

Despite these hours in bed and several hours more a day gambling, he accomplished a prodigious amount of work by using every spare moment. Wasting time—especially on friends—he considered "an abomination." With his well-spent time, he soon grew to be Europe's second-best doctor. (Vesalius was considered the first.) Great men tried to lure him away from Milan with great promises. The King of Denmark offered him a fabulous salary as court physician, but Cardano refused to go to a cold, non-Catholic country. "I turned down several even more generous offers," he notes, "one from the King of France . . . and one from the Queen of Scotland who tried to get me with lavish promises."

Only once did he leave Italy on a medical mission and that was in 1552 when he went to Edinburgh to treat John Hamilton, the Archbishop of Scotland. Along the way he stopped off to sightsee and receive the accolades of royalty and prominent scientists. In France he noted that he visited the King's private treasure vaults where he saw "the perfect horn of a unicorn." "Everywhere I was given the same sort of reception that Plato had received long ago at the Olympic games," he reported, with his characteristic penchant for comparing himself to famous figures of antiquity.

In Scotland he administered the usual brews and medicines to his patient who was suffering from what had been diagnosed as either consumption or asthma—neither of which any doctor could cure. But Cardano was far ahead of the rest of the medical world—and even by today's standards his treatment of the case would inspire respect. Apparently he suspected, after weeks of observation, that the Archbishop was suffering from an allergy, for he ordered him not to sleep on a feather pillow. The Archbishop followed the doctor's orders and his symptoms disappeared—to the relief of both himself and Cardano.

From Scotland Cardano made his way home, stopping off in England to have an audience with young King Edward VI, for whom he cast a horoscope predicting a long life marred by a few minor ailments. Cardano just made it back to Milan before Edward died.

Cardano was now a man of fifty-four and at the height of his career. He had money, fame, position, and three healthy children. His wife Lucia had died seven years previously and his grief had been short-lived. He was truly one of the most successful men of the Renaissance, having risen from poverty to a position where he and his work were crowned with glory. He himself was not unaware or overly modest in the face of fame, and in his biography devotes a whole separate chapter to listing approximately seventy-five books in which he is mentioned with praise. Then he appends a very short list of eight authors who have damned him—including Tartaglia—and remarks that "not one of these men has any learning beyond grammar school." In his constant comparisons between himself and great ancient figures, he notes that neither Galen nor Aristotle had as many commendary references made about them in their lifetimes, completely ignoring the fact that books were scarcer then because the printing press had not been invented.

Then, in the midst of all his good fortune, Cardano says that his house shook from basement to attic on the night of December 20, 1557, as though an earthquake had passed. Here, indeed, was a terrible omen. The next day, while he was eating with his older son, Giambatista, a friend stopped by and mentioned that Giambatista was going to be married. Cardano knew nothing about it. Giambatista, in an astonished voice, said that this was the first he too had heard of a wedding. But a couple of days later a messenger rushed up to the Cardano house with the information that Giambatista had, indeed, married the daughter of a down-and-out spendthrift. Her reputation for free and easy ways was known throughout town, and even Cardano must have heard of it, for when Giambatista brought his new bride home, the door was slammed in their faces.

Cardano had had great hopes for his older and favorite son, hopes that Giambatista had wrecked by marrying a trollop with three sisters and a mother to support. Cardano obviously had expected Giambatista to marry into a family of good background and wealth. After all, should not the son of Italy's most prominent and famous doctor be able to take his pick of young, available ladies? The answer would be "yes" except for the fact that Giambatista was humpbacked—a deformity that had affected his grandfather to such a slight degree that it had been passed off as round shoulders but which was much more pronounced in Giambatista. Cardano, the doting father, was blind to the physical handicaps of his beloved son. Even as far as the boy's abilities went he was slightly myopic, believing that Giambatista was a near genius. And as for the foolish marriage, he was certain that the girl and her parents had driven his son into it in order to lay their hands on some of the Cardano fortune. It never occurred to him that his son might really be in love.

Driven from home, Giambatista moved in with his wife's relatives and was soon saddled with their support. Cardano sent him money constantly and even set aside for him the fees from a certain part of his practice. Along with the money, he sent assorted bits of advice and cheer, taking care to point out that "from the very beginning I had to support my family myself," and neglecting to mention that he had not been burdened with the support of a flock of in-laws. He sent his son two books he had written, *Consolation* and *Adversity*, sixteenth-century counterparts of our own positive-think bestsellers, along with the comforting thought that in spite of poverty and misery Giambatista could at least be thankful that he had "a father through whose work your name will endure for many ages." Apparently Renaissance men were not much different from their medieval predecessors when it came to balancing immortality against mortal happiness.

The wasp that had circled Giambatista on his christening day had, in Cardano's eyes, been an omen of evil. A few days before the wedding even the servants had felt the house shake—another evil omen. And then, about two months after Giambatista's second wedding anniversary, another omen appeared in the form of a red spot on Cardano's finger. That very day a messenger arrived with the news that Giambatista's wife was dead and that Giambatista had been arrested for her murder.

At the trial the dreadful story was made public of how Giambatista had been taunted by his wife and mother-in-law who said that his two children were not really his. Almost insane with rage and jealousy, Giambatista had paid a servant to purchase some arsenic and mix it in his wife's food. The arsenic was bought, but before it could be used Giambatista was overcome with remorse and horror at the idea, or so he said. A few days later— whether at his instruction or not is unclear—the arsenic was mixed into a cake and served up to the whole family. Giambatista himself took a piece and became ill, as did his mother and father-in-law. His wife, Brandonia, already weakened by the birth of her second child, had a piece, became dangerously ill, and a few days later died.

Cardano did everything he could for his son but refused to visit him in his cell. He paid all the expenses, testified as to the "worthless, shameless woman" his son had married, and hired the best lawyers. Five doctors were brought in who stated that Brandonia had not been poisoned—or at least, had not received a fatal dose. It seemed that a not-guilty verdict would be handed down when for some inexplicable reason, Giambatista confessed. He was sentenced to death and shortly after was tortured and beheaded.

The death of his son marked the second turning point in Cardano's life. He never recovered from his grief and even years later was unable to write more than a few pages without turning the subject to his dead son.

Everywhere he looked in Milan there was something that reminded him of Giambatista, something that only intensified his grief. And so he gave up his lucrative medical practice, took his grandchild with him, and moved to Pavia. Yet even there he was unable to forget "the great disaster." His only consolation was his baby grandson, Fazio, whom he adored, despite the lingering doubt about the child's paternity. His granddaughter had died a few days after her father's execution; and his own daughter, Chiara, was childless although married. Aldo, his younger son, was a good-for-nothing who sponged off his father and did nothing but gamble. And so Cardano pinned all his hopes on little Fazio.

There is evidence that the almost unbearable tragedy affected Cardano's mind, for beginning in Pavia he reports a number of intrigues and attempts on his life. Even the execution of his son, he says, was done "in the hope that I would die or go insane from grief." He spent only a few years in Pavia before accepting a position at the University of Bologna, and in those few years was plagued by the "calumny, defamations and treachery of my dishonest accusers." He recounts how his enemies "plotted to murder me" by dropping beams and stones on him or bribing his servants to poison him. Every loose tile on a roof, he was sure, had been rigged by his enemies to fall on him. His servants could not spend an evening out but that Cardano did not suspect some plot to murder him. Complicated intrigues to besmirch his name supposedly extended even to Bologna, where his enemies accused him of professional incompetence, sexual perversion, and other sundries. All these plots, Cardano assures his readers, were motivated by jealousy. A more realistic explanation would be that they never happened at all but were merely creations of the poor man's mind.

In time, the really pathological suspicions passed, but to his dying day, a motif of persecution was present. He was forever listing the "many troubles and obstacles in my life"—foremost of which was the death of Giambatista. The bitterness of that never left him. It was a great miscarriage of justice that had robbed him of his favorite, and as proof, he cited the fact that everyone connected with the prosecution of his son had been afterward struck with "some dire mishap"—untimely death, imprisonment, bereavement, or disgrace.

Cardano's constant harping on his troubles, however, is not without a note of pride. Often it is not pity he seeks to inspire, but admiration. "My struggles, my worries, my bitter grief, my errors, my insomnia, intestinal troubles, asthma, skin ailments and even phtheiriasis, the weak ways of my grandson, the sins of my own son . . . not to mention my daughter's barrenness, the drawn-out struggle with the College of Physicians, the constant intrigues, the slanders, poor health, no true friends" and "so many plots against me, so many tricks to trip me up, the thieving of my maids, drunken coachmen, the whole dishonest, cowardly, traitorous, arrogant crew that it has been my misfortune to deal with"—a lesser man would certainly have fallen under the burden of these problems, but not Girolamo Cardano.

It was true that he did have phenomenally bad luck in raising his children and grandson—all of whom turned out badly except his daughter. His only complaint in that quarter was that she had no children, and more than anything, Cardano wanted heirs. He, himself, realized that he was partly to blame for his sons' poor characters. He had surrounded the motherless children with his rowdy gambling companions and ignored their upbringing until it was too late.

In 1562, when Cardano went to Bologna to teach at Europe's oldest medical school, he took his son Aldo along, but after a short while found it impossible to live with the boy, who did nothing but drink and gamble. Aldo got a place of his own and soon found himself so deeply in debt that he resorted to stealing—not that this was the first time he had turned thief to pay his way. He broke into his father's house, took some money and jewels, and was caught. Disgusted, Cardano had him banished from the area; he was tired of bailing Aldo out of the staggering debts he always managed to run up.

With Aldo out of the way, life went along peacefully. Cardano continued to pour out books at a phenomenal rate—eighteen when he was past seventy. The disgrace of his sons, while it had not left him, was at least behind him. Life had resumed its old order when suddenly without warning—not even a bump or rattle—Cardano was arrested and thrown into prison. An old man, Lear-like, with sixty-nine years on his shoulders, he had somehow, inadvertently, committed heresy. After three months in prison, he was allowed to recant privately, sworn to secrecy about the whole proceeding, and released.

No one knows exactly what heretical act he was charged with—perhaps dealing in black magic to effect cures, for in his auto-biography, written later, he makes a special point of stating that he never had any dealings with necromancy or magic and credits "divine assistance," not his own ability, for his success in medi-cine. Certainly, it is not like Cardano to let others take the credit for his achievements—even if the "others" be God. It is more than likely that he did it with one eye on the Inquisition.

After being released from prison, Cardano went to Rome where one of his influential friends miraculously persuaded the Pope to grant the ex-heretic a pension. Cardano stayed in Rome until his death, writing, studying, and casting horoscopes for the Papal court. According to legend, he forecast the day of his own death—September 20, 1576, four days before his seventy-fifth birthday—and when the day arrived and he was still very much alive, committed suicide to fulfill the prediction. This is the kind of colorful end one would like to believe, and it is quite likely that it was tacked on to the already colorful story of Cardano's life for just that reason—color.

His stormy life over, his reputation lived on for at least another century. The name Cardano which he had toiled so hard to im-mortalize is little known today, but his mathematical bequest has produced a progeny beyond anything he could have dreamed or read in the stars.

René Descartes
1596-1650

Truth lay just around the corner like a veiled statue waiting for men to uncover it. The causes and effects of everything from comets to heart palpitations could be made known, and, in Descartes' words, men would soon be "the lords and possessors of nature."

This was the epic scale on which Descartes dreamed. He had discovered a wonderful new method for discerning the truth—a method totally different from the ones men had been using. It was to apply the method of mathematics to every area of life. By using logic, step by step, all the secrets of nature could be laid bare. And with knowledge would come control.

Today, three hundred years later, when men can cause rain to fall, plants to grow, or fatal diseases to disappear, the idea of mastering nature or of understanding the universe—from the molecule to the moon—does not seem so new and wonderful. But three hundred years ago, the idea that man might actually control nature was either unimaginable or sacrilegious or both. It was arrogating God's power. Yet Descartes, with a handful of other men—Francis Bacon being the most famous—dared to think along this truly original line. And slowly, bit by bit, man *has* made himself lord and possessor of nature.

Descartes never envisioned how difficult and perhaps never-ending the task was. He expected to discover the complete truth in his own lifetime—or at the latest, believed it would be found by the next generation. Today we know that mastery is much more difficult and that the whole truth may never be discovered. Yet Descartes optimistically advised his ailing friends to hold on a little longer—cures for disease and perhaps the secret of eternal life were soon to be found. And it was man's mind—not miracles or magic—that would find them. Thus, he ushered in an Age of Reason and helped set the stage for the modern world.

Born on March 21, 1596, twenty years after the death of Cardano, René Descartes came of a noble French family, "one of the best in all Tourain." His mother died a few days after his birth and the weak, sickly baby, too, was "condemned to dye young" by the doctors. But he managed to survive and grew into a pale, serious child with the same dry cough that his mother had had—perhaps an indication of tuberculosis.

Descartes had an older brother and sister, and after his father remarried, a younger brother and sister. But his favorite companions were his nurse and an unknown little girl who squinted. Throughout his life he had few close friends—perhaps not more than half a dozen, but to them he displayed a remarkable degree of devotion and loyalty. When he came into his inheritance, he gave his nurse a generous pension and for the rest of his life was partial to women who squinted.

A quiet, studious child, he was given the prophetic nickname of "Philosopher" by his father. At the age of ten the serious little boy was sent away to La Flèche, a new Jesuit school that came to be one of the most celebrated in Europe. There Descartes easily breezed through the regular courses in logic, ethics, metaphysics, literature, history, and science, and then went on to do independent work in algebra and geometry—which was to bear fruit later. He was a voracious reader and, by his own admission, read "all the books that had fallen into my hands, treating of such branches as are esteemed the most curious and rare."

But after eight years of intensive study he declared that the one thing he was surest of was his ignorance: "As soon as I had finished the entire course of study, I found myself involved in so many doubts and errors, that I was convinced I had advanced no farther . . . than the discovery . . . of my own ignorance." Philosophy, especially came in for special scorn. "I saw that it had been cultivated for many ages by the most distinguished men, and that yet there is not a single matter within its sphere which is not still in dispute." Systems of ethics were "towering and magnificent palaces with no better foundation than sand and mud." Mathematics alone "especially delighted . . . on account of the certitude and evidence of their reasonings."

And so, at the age of eighteen, he returned home, convinced that almost everything he had learned was either useless or riddled with error. He spent a few months visiting his family, took up fencing and riding, and was then packed off to the University of Poitiers for two more years of study—this time law. Descartes' father was a lawyer and apparently planned to have his son follow the same profession. But as soon as Descartes received his degree, he "entirely abandoned the study of letters, and resolved no longer to seek any other science than the knowledge of myself, or of the great book of the world." His plan was to travel along "the right path in life"—which usually means the straight and narrow and makes for very dull reading. Fortunately, this life of reason that Descartes planned had several romantic detours.

He set out for Paris where he amused himself for a few months gambling and, with proverbial beginner's luck, won consistently. When this life of "lesser debauchery" palled, he simply gave it up and moved to another part of Paris "without so much as giving his Friends or Kinsmen notice thereof." Social life and all that went with it meant very little to him. He tired of it easily and retreated into a shell of scholarship. Despite his resolution that he would have no more to do with books, he spent the next two years "over Head and Ears in the Study of mathematicks," quietly working in seclusion and lying half the day in bed thinking. Spending long hours in bed was a habit which he had acquired in school through the indulgence of one of his teachers and kept for the rest of his life. He never got up before eleven; sometimes not before noon.

This quiet life was brought to an end when he accidentally met an old friend on the street who then began visiting regularly and dragged Descartes back again "upon the Stage of the World." Shortly after, Descartes retreated once more, this time by joining the army, which would also give him a look at another side of life. Just past his twenty-second birthday he volunteered for the army of the Prince of Nassau and was sent to the little city of Breda in Holland.

Even in the army Descartes was lured back to scholarship. One day—on November 10, 1618, to be exact—he was wandering through the streets when he saw a crowd gathered before a posted notice. Curious, he ambled over and asked someone to translate the Flemish into Latin or French for him. An older man read off the mathematical challenge contained in the notice and also the problem to be solved. Descartes remarked that it was an easy problem and his translator—Isaac Beeckman, who happened to be one of Holland's greatest mathematicians and doctors—proposed that he solve it if it was so easy. Descartes did, and Beeckman, realizing that here was no ordinary gentleman-soldier, became his friend and mentor.

He urged Descartes to continue with mathematics, setting him several original problems to work on, including finding the law for the velocity of falling bodies. Neither of them knew that Galileo had already found the law—$32t$ feet per second per second, where t is the number of seconds—for scientific communication was slow and inefficient.

The influence of Beeckman on Descartes was tremendous, and Descartes readily conceded that Beeckman was "the inspiration and spiritual father of my studies." "I slept and you awakened me," he said—awakened him to pioneer a vast new field of mathematics.

Four months after the fateful meeting, Descartes reported to Beeckman his discovery of a new way of studying geometry. Descartes had been disturbed by the methods of Greek geometers for a long time. There seemed to be no underlying system of attack in their ingenious proofs which, Descartes said, "exercize the understanding only on condition of greatly fatiguing the imagination." He proposed to correct this by handling lines and plane figures on a graph. The graph is made by marking off units on a horizontal line, the x axis, and a vertical line, the y axis. On the x axis, any point to the right of the xy intersection is positive and any point to the left is negative. Similarly, on the y axis, any point above the intersection is positive; and below, negative (Diagram 4). Figures or lines can be drawn on the graph and, according to their position, described by numbers. For instance, any point can be described by two numbers, one representing a distance along the x axis and the other along the y axis. In Diagram 4, point P is represented by the numbers 1 and 2.

All the rules of Euclidean geometry hold true in this new coordinate geometry. For instance, in Euclidean geometry a line is described by two points, and in Descartes' geometry the same is true. Since a point in Cartesian geometry is described by a pair of numbers, then a line can be described by two pairs. In Diagram 5, the line PP' is described by $(1, 2)$ and $(3, 4)$ (the first number in a pair usually denotes the number along the x axis and the second along the y).

One advantage that Descartes' geometry has over Euclidean, is that the length of a straight line can be easily determined and expressed by a number. His lines have a definite magnitude—they are not simply twice or half the length of another line. For instance, the line in the above diagram can be taken as the hypotenuse of a right isosceles triangle whose arms are each 2. There-

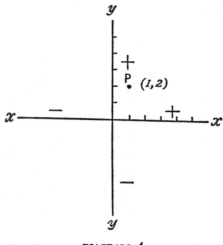

DIAGRAM 4

fore, according to the Pythagorean theorem, the line *PP′* is $2\sqrt{2}$ in length.

From Euclidean geometry we know that it is possible to extend a line to infinity and also that even the shortest line is made up of an infinite number of points. Therefore, in analytic or coordinate geometry, it should be possible to describe a line, or a

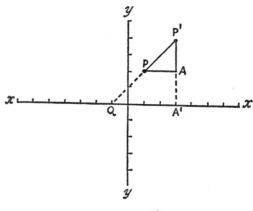

DIAGRAM 5

line extended, by any two of these infinite points. Can this be done without a trial-and-error test of all the possible pairs of numbers, and if so, how? The solution lies in finding whether some sort of relationship exists between the values of x and y for any point on the line and then expressing this relationship in an equation.

From the triangle PAP' (Diagram 5), a similar right triangle $QA'P'$ can be constructed by using the x axis as one arm, extending $P'A$ to make the second arm, and extending the hypotenuse $P'P$ to the x axis. We know from Euclidean geometry that the ratio between the sides of similar triangles always remains the same, no matter how great or small the difference in size between the triangles. Therefore, in coordinate geometry, too, the ratio of the sides, i.e., the ratio of the x and y values, will remain constant. And since x and y determine the points on the line QP', there will be a constant ratio between the pairs of numbers on the line. All that remains is to express this relationship in an equation.*

The arms of the new right triangle $QA'P'$ are each 4 units long. The arm lying on the x axis extends across the xy intersection into the negative side so that one of its four units is a minus unit. The arm is still, however, four units long. The minus unit here simply means that the arm on the x axis starts before the arm on the y axis, making it always register 1 unit less than the line determined by the y axis. It is as though two runners, both going at the same constant speed, ran a race 100 yards long but one runner started out 10 yards ahead of the other. At the end they would each have run 100 yards, although one man would be 10 yards ahead. So it is here; y is always 1 unit ahead of x. Therefore, $y = x + 1$, or to put the equation in the more general form, $x - y + 1 = 0$. Any pair of numbers describing the line

---

* So as not to confuse the reader with too many technical details, a line making a 45° angle with the axis was chosen, thus eliminating the necessity of discussing slope. The slope of the line under consideration is 1 and therefore does not alter the calculations.

PP' or *PP'* extended must fit this equation. By choosing any value we wish for x and substituting it in the equation we can find the corresponding value for y.

By plotting a few more lines, Descartes discovered that the ratios between the points determining a straight line all give rise to the same general equation form: $ax + by + c = 0$, where a, b, and c are constants. Thus, all straight lines can be described by one single type of algebraic equation.

So far the applications of Descartes' analytic geometry may seem to shed little new light on the subject of lines. It is simply another and possibly more complicated way of looking at the subject. But in addition to points and lines, Descartes studied circles and other curves, and here analytic geometry really proved itself of use. The Greek constructions of curves—aside from the circle—were difficult and complicated. The relationships between different kinds of curves were obscure and almost impossible to fathom. For instance, what do the curves, a circle and a parabola, have in common outside of their not being straight lines?

When Descartes studied curves on his graphs, he discovered some surprising results. He noticed that all curves fall into certain general categories of equations, just as straight lines do. A circle, for instance, with its center at the intersection of the x and y axes (Diagram 6), has a circumference that is everywhere equidistant from O. Only points lying on this circumference describe the circle. The radius from O to the circumference remains constant—that is, its length is always the same; only the values of x and y change. Using the Pythagorean theorem again, with the radius, r, as the hypotenuse of a right triangle and x and y as the arms, an equation can be formed. In this case the equation is $x^2 + y^2 = r^2$. The variables x and y can be any numbers we want, just as long as their squares add up to r^2. Thus, the general equation for a circle with its center at O always follows the form $x^2 + y^2 = c^2$.

Other curves—ellipses, hyperbolas, parabolas—all yield general equations of the second degree, whereas the equation for the straight line is of the first degree.

An analogy of two runners was used in describing the relation between x and y for a straight line. One runner is running along the x axis and the other along the y. In Diagram 5, they were both running at the same constant speed and the line QP' described

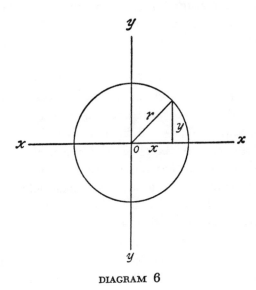

DIAGRAM 6

their relation to each other. It is as though the line were the scorekeeper for the two runners, letting each know how far the other has run. If one runner had been going faster than the other, the line would still be straight but would slant at a different angle.

This same analogy of runners and scorekeeper can be used in describing curves. In this instance, however, the runners are not both going at *constant* speeds. Sometimes x is speeding up while y is slowing down or running at a constant speed. A curved line rather than a straight would depict a race of this type.

Thus, analytic geometry contains the implicit idea of lines as moving points. The scorekeeper or line moves along to show the relation between x and y at different points in the race. Euclidean geometry never showed this!

It was the news of this new kind of geometry—or a new way of handling an old kind of geometry—that Descartes excitedly conveyed to Beeckman. He had found a unifying principle behind the complicated constructions of geometric figures, namely that for each type of line there is a unique equation form that describes it. Descartes still had to work out other details and applications— but that came later.

He had not forgotten that he had joined the army to see the world and "to Study the different natures of men." Holland began to bore him. It was too quiet and peaceful, besides, he "really did take delight in War"—a delight his biographer, Baillet, dismisses as being due to "the effect of the heat of his Liver." Descartes decided to go to Germany where the Thirty Years War was being fought. Uncertain what army he should join, he finally volunteered for the Duke of Bavaria's forces "without knowing precisely against what enemy they were to march."

The army settled down in its winter quarters in October, leaving Descartes little to do but sleep, eat, and work on his geometry. A month later, on November 10, 1619—exactly a year after he had met Beeckman—he had three remarkable dreams, one right after the other. The dreams occurred while he was sleeping in a stove —in those days any small room used for heating was called a stove. While asleep in such a room, Descartes dreamed his three dreams.

In the first, strong winds were blowing him away from a church and toward a group of people who were not at all disturbed by the force of the gale. He then woke up, fell asleep again, and had another dream about a storm in which a great clap of thunder filled his room with fiery sparks. Terrified, he woke up again and soon fell asleep to dream his third dream in which he was turning the pages of a dictionary and found himself holding a sheaf of papers, one of which contained a poem beginning *"Quod vitae sectabor iter?"*—What life shall I follow?

Descartes interpreted these dreams to mean that he had been wandering about in the world, buffeted by forces he could not control, and that he now had to choose what sort of life he should follow. The clap of thunder was a warning to him to choose before it was too late, and the dictionary seemed to indicate that he should pursue knowledge. The search for truth should be his career.

He was well on his way in his search for truth and that very day had written in his diary, "I begin to understand the foundation of a wonderful discovery . . . All the sciences are interconnected as by a chain; no one of them can be completely grasped without taking in the whole encyclopaedia all at once." What was this "wonderful discovery"? No one knows for sure. Until recently there was considerable agreement that it was his analytic geometry, but the discovery of new material rules that out. It may have been some further development of his "method" for finding the truth, or it may have been the discovery of another application of his geometry.

At any rate, fired with new zeal, he continued to develop his mathematics and philosophy. The suspicion that all knowledge is connected in the same way as geometric theorems set him off on the mammoth task of finding these connections or chains of reasoning. His approach to this task incorporated the methods of mathematics and was based on the following rules:

1. Never to accept anything for true which I do not clearly know to be such.
2. Divide each of the difficulties under examination into as many parts as possible.
3. Begin with the simplest and easiest and then work step by step to the more complex.
4. Make enumerations so complete and reviews so general, that I might be assured that nothing is omitted.

Here was the method of the geometer which Descartes planned to apply to all areas of life.

These four rules were the foundation on which he constructed all his work, philosophical as well as scientific. Rejecting everything he knew, he erected the famous proof that he himself existed ("I think, therefore I am") and finally that God existed.

The importance to us of his method, however, is not in religion, but in science, where "the long chains of simple and easy reasonings by means of which geometers are accustomed to reach the conclusions of their most difficult demonstrations" can be used to discover "all things." Like the Greeks, Descartes believed the world to be a mathematical machine in which "there is nothing so far removed from us as to be beyond our reach, or so hidden that we cannot discover it."

By using "simple reasonings," Descartes hoped to uncover universal laws—to lay bare all the secrets of a mechanical universe. This ancient Greek view of the universe appeared startling in its new dress. Plato had said that the universe can be explained mathematically because "God ever geometrizes." Descartes added a new frill—that with understanding will come control. The presupposition of a mechanical universe is superficially innocent. But the idea of understanding and controlling that universe is downright heretical. The Bible says that God made the universe; it does not specify whether or not He made it in a mathematical mold. But the Bible is specific in its injunctions not to try to understand God's ways, much less arrogate God's power or control of the universe. Fortunately for Descartes, few people realized the essentially heretical nature of his ideas until it was too late. Even he himself was unaware that he was sabotaging the Church.

Descartes stayed in the army for only about a year and a half longer, quitting when his general, who had been deserted on the battlefield, was killed. He had no intention of going back to Paris, which was in the grip of a plague, for he still had much to read of "the Great Book of the World." The next chapters took him traveling through Hungary, Moravia, Silesia, Poland, and the provinces of Germany, where he occupied himself in "holding intercourse with men of different dispositions and ranks, in collecting varied experience, in proving myself in the different situations into which fortune threw me, and, above all, in making such reflection on the matter of my experience as to secure my improvement."

His well-laid plans almost came to an unexpected end during a trip on the North Sea when some unprincipled sailors were overcome with the idea that they had a wealthy passenger aboard and decided "to knock him on the Head, to fling him into the water, and divide the Spoil." But Descartes got wind of what they were up to, and turning his knowledge of human nature to good use, took the offensive. He drew out his sword, which no gentleman ever went without, and threatened "to run them through if they durst but hold up a finger against him." The intimidated sailors brought him to shore safely with all his possessions intact.

After his wanderings around Europe, Descartes returned to Paris and moved in with a distant relative to live "the most plain, simple kind of life." A plain, simple life in those days consisted of having a few servants and dressing in the typical gentleman finery of "a Plume of Feathers, Scarf and Rapier."

Descartes had seen all he needed of the world; now his real work was before him. He had at last decided against a profession, feeling that a life of pure reason was his forte. His reputation in Paris as a thinker and philosopher spread so fast that he was plagued with constant visitors. Even perfect strangers "were forward to slip in amongst the rest" at his house, and publishers plied him with gifts in order to get any books he might write. To a man who had spent the last few years wandering about by himself, the sudden deluge of people was "very tedious and burthensome." Partly because he could find neither the time nor solitude to think, and partly because of urgings by respected acquaintances not to

deprive mankind "of the benefit of his Meditations," he decided
to leave Paris and settle in peaceful little Holland.

He left Paris for the country in order "to accustom himself by
degrees to Cold and Solitude" and then in the spring of 1629
moved to Holland, away from "the heat of the climat and croud
and hurly-burly of People," keeping in touch with Paris through
an old friend from school days, Father Mersenne. Mersenne, who
was said to have more in his head than all the universities together, acted as a clearinghouse for the various scientific works of
his day. He took the place of the still nonexistent scientific journals, broadcasting through letters news of the latest achievements
by Pascal, Descartes, Roberval, and others.

To say that Descartes "settled" in Holland is perhaps a misuse
of the word. All in all, he lived there more than twenty years—
with time out for three quick trips to Paris—but constantly moved
about like "the Israelites in Arabia." He seldom stayed in cities,
preferring the suburbs or "lonesome Houses far in the Country."
(Ironically, one of the homes of this rational philosopher was
later converted into an insane asylum.) He took precautions to
hide his whereabouts from friends by having all his mail sent to
another address. Nevertheless, even in Holland his fame "followed
him as a Shadow," so that on an average of once a year he was
forced to pick up and move elsewhere to avoid visitors.

There was so much to do and if he wanted to get it done, he
could not squander his precious time with friends, although he
was willing to admit that "the greatest good in life is friendship."
A strange admission for a man as antisocial as Descartes.

His work, rather than people, was his companion. He was an
observer on the stage of life—not one of its actors. "I stroll every
day," he wrote a friend, "amid the Babel of a great thoroughfare
with as much freedom as you could find in your garden paths
and observe the people I pass just as I should the trees in the
forests . . . The noise of their bustle does not disturb my re-
flections more than the babbling of a stream."

And so in Holland he went his solitary way, already launched on his great work. He had perfected his "method," which is described in *Rules for the Direction of the Mind,* by the time he moved to Holland, although the *Rules* was not published until after his death. He now began the *Discourse on Method,* which also contains, in addition to his method, three essays, *Meteors, Dioptrics,* and *Geometry.*

Long before the *Discourse,* the most famous of all his books, was ready for publication, he dropped it to write a philosophical treatise entitled *Le Monde,* which covered everything from the stars and Moon to human anatomy. Apparently Descartes felt that the book would cause a sensation and might give offense to certain members of the Church. The Jesuits had been kind to him; he had great respect for them and he wanted their respect in return. Therefore, he decided not to sign his name to the book. "Now that I am to be not only a spectator of the world but am to appear an actor on its stage," he wrote, "I wear a mask." True, he was completely indifferent to any fame he might gain as an author, but in this case his desire for anonymity stemmed from other considerations. He was a timid man who had never done well in social situations. Although he could scare up enough bravado to save himself from thieving sailors, he could not summon up enough courage to destroy himself in the eyes of his friends.

No sooner had he sent his book off to the printer unsigned so that "I may always be able to disavow it," when he learned that "persons"—meaning the Inquisition—"had condemned a certain doctrine in physics"—meaning the Copernican theory—"published by another individual"—meaning Galileo. Despite his violent passion for truth, Descartes admitted that these "persons" had as much authority over his actions as "my own reason over my thoughts." And this, plus "other considerations," was "sufficient to make me alter my purpose of publishing." After all, he reasoned, since he himself could see nothing sacrilegious in Galileo's writings, perhaps "among my own doctrines likewise some one might

be found in which I had departed from the truth." The day before the book was scheduled to go to press he wrote Father Mersenne, "You will only have a better opinion of me when you learn
that I have definitely decided to suppress the Treatise and so lose
the results of four years' work in order to submit with complete
obedience to the Church."

Descartes' position becomes even more untenable when it is
considered that he was living in Holland, a country where religious
tolerance was at an all-time high and where "everybody is for the
movement of the earth." Therefore, it was not his skin he had to
save, but his face. There was little chance that he, like Galileo,
would be thrown in prison. All that he had to worry about was
that if his book were banned by the Church, it would be impossible for him to hide behind a pseudonym. All his Jesuit teachers
would find out what their star pupil had done and condemn him.
Some people believed that Descartes' desire for approval drove
him not only to suppression of certain portions of his work, but
to actual distortion. Hobbes, for instance, maintained that Descartes did not believe everything he wrote—especially in religious
matters but did it "meerly to putt a compliment on the Jesuites."
Apparently Hobbes, being somewhat atheistic himself, found it
difficult to believe that other intelligent men could believe in God.

While *Le Monde* was making its rocky way to oblivion, Descartes was hard at work scattering his genius over several fields of
science—physics, chemistry, medicine, mathematics, and astronomy. Thomas Hobbes regretted that Descartes did not concen-
trate solely on mathematics and "was wont to say that had M.
Des Cartes kept himselfe to geometrie, he had been the best
geometer in the world."

Nevertheless, some of Descartes' work in other fields did bear
fruit. In medicine, for instance, he was one of the first to accept
and promote Harvey's theory of the circulation of the blood.
Harvey, an English doctor, had fared so poorly "after his booke
of the Circulation of the Blood came out, that he fell mightily in

his practize, and that 'twas beleeved by the vulgar that he was crack-brained; and all the physitians were against his opinion." Descartes himself had discovered that blood circulates and immediately came to Harvey's defense. Having a man of Descartes' stature as supporter did much to gain acceptance for the theory.

Descartes had taught himself medicine by going "almost every day to a Butcher's Shop to see him kill Beasts, and from thence he caused the Parts of such Animals to be brought home to his Lodging, as he design'd to Anatomize at his own leisure." In medicine, as in every other science, he wanted a body of knowledge grounded upon infallible demonstration. According to his famous method, these demonstrations were supposed to be based on logic and self-evident axioms similar to those used in geometry. But anatomy is not mathematics. It cannot be conceived in a vacuum. It must proceed from observed facts, and for this reason Descartes was driven to experiment. Only from observation and experiment can the natural and physical sciences gather the necessary data to create laws. For instance, no amount of juggling numbers will reveal what man's normal temperature is. The only way to find out is to take the average of the temperature of a number of healthy people, which then yields the medical truth that the normal temperature is 98.6°. The Greeks in their worship of Reason, scorned such a thing as experimentation. Reason, not man's senses, was the only valid instrument. This view had led to many highly original and fascinating discoveries, such as that men have more teeth than women, a 10-pound weight falls twice as fast as a 5-pound, etc.

Descartes, however, apparently did not distinguish clearly between the two ways of discovering truth—pure reason and experimentation. His one basic premise in medicine was that the bodies of men are mechanical and subject to exact laws. But this premise in itself cannot be used to reason out a body of medical knowledge; therefore he freely used experiments, although his "method" would seem to exclude them.

One of his most important contributions to medicine was his creation of a hypothetical body. This hypothetical body was the idealization of the human body, just as the forms studied in geometry are idealized forms that do not exist in nature. By assuming a hypothetical ideal it is possible to learn more about the actual than if the actual itself be studied. Today almost every branch of science uses the concept of an ideal which does not really exist in nature but which gives better results than the aberrations of all the imperfect actualities. For instance, to measure the speed of a falling object, all extraneous influences, such as friction and wind, must be discounted. Yet where is friction absent except in a vacuum, and where is the perfect vacuum except in the minds of men? Galileo had to formulate his law by idealizing the results of his experiment. This revolutionary concept of an ideal in experimental science was borrowed from the ideals of geometry.

Descartes had tremendous expectations for medicine: life eternal, perfect health, the conquest of all disease—these were the goals at which he aimed. His great expectations dwindled, however, as he confronted the magnitude of the subject. At first he had hoped that his studies would enable him to prolong life indefinitely; later he was more modest and hoped only to "obtain some delay of Nature"; and finally, nothing more than to "retard grey hairs which began to grow upon him."

Descartes' affinity for medicine may have arisen from his sickliness as a child, rather than from intellectual curiosity. Throughout his life he was overly concerned with physical health and, next to virtue, considered a healthy body to be the greatest asset on earth. From the age of nineteen on, he doctored himself according to his own ideas, which included a "spare regular Diet and moderation in his Exercize." He pampered himself by never getting up until he felt like it and believed that a person should alway keep something in his stomach, especially "Roots and Fruits, which he believed more proper to prolong the Life of Man, than the Flesh of Animals."

Although Descartes never reached his ultimate goals in medicine, he did make worthwhile contributions. In other fields he was not as successful. The Cartesian ideas in astronomy are best forgotten, and much of his work in physics is, in the idiom of the day, "crack-brained." His error usually lay in first formulating theories through pure reason and then trying to make the observable facts fit the theories.

But in geometry, he had no peer. If the aim of modern mathematics is to simplify and consolidate mathematical theories, Descartes could be called the first of modern mathematicians. Cartesian geometry united geometry and algebra, two fields that had until then been thought to be separate. That a subject that deals with forms and another that deals with symbols could have anything in common seems far-fetched, yet in analytic geometry they are welded together.

One or two examples will show how close—or even identical—geometry and algebra are. It has already been shown how lines can be mapped or plotted on a graph and then described purely in terms of numbers. A straight line gives rise to a first-degree equation; a circle and other conic sections yield second-degree equations; and the more complex curves yield equations of higher degrees. Now if lines give rise to equations, the reverse may also be true. Consider a simple first-degree equation in two unknowns, such as $4x - y = 1$. Such an equation has an infinite number of solutions. Values of x and y respectively can be 1 and 3, ¾ and 2, ½ and 1, etc. The equation expresses a relationship between x and y that is constant even though the values of x and y may change. x is always one less than four times y.

This relationship can be expressed on a graph. That is, by plotting any two values of x and y on the graph, we get two points which determine a line (Diagram 7). It is unnecessary to plot more

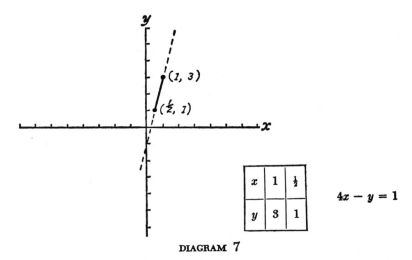

DIAGRAM 7

than two, for an equation of this type always describes a straight line and any points lying on this line will satisfy the equation.

In the previous chapter it was pointed out that to solve equations in two unknowns with one definite set of answers instead of an infinite number of sets, we must have two equations, such as the one above, $4x - y = 1$, and another, such as $x + 2y = 7$. Plotting the second equation on the same graph as the first, we find that the lines cross at the point where $x = 1$ and $y = 3$ (Diagram 8). If the two equations are solved algebraically, the roots are $x = 1$ and $y = 3$. Thus, sets of algebraic equations can be transformed into lines and solved by taking the point of intersection as the roots.

The connection between geometry and algebra should now be obvious. They are not only similar, they are Siamese twins joined together by analytic geometry! From a geometric point of view, analytic geometry presents points on a line, points in a plane, and points in space by giving these points a numerical value. Points are simply reduced to numbers. From an algebraic point of view, analytic geometry converts numbers into points. These points determine a line, a plane, or a solid and enable mathematicians to visualize equations. Thus geometry is algebrized and algebra is geometrized.

Analytic geometry did more than to consolidate two utterly different branches of mathematics. It opened the way to the invention of calculus by proving a graphical or visual way of studying motion. Analogies of two runners were presented previously

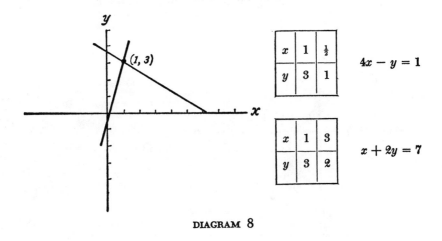

DIAGRAM 8

to show how Descartes' lines can represent motion. The importance of these analogies will become more apparent in the chapter on Newton.

One less evident aspect of coordinate geometry was missed by Descartes, which is perhaps fortunate, for he might have thrown the whole thing away if he had noticed it. Coordinate geometry assumes that for every point on the x or y axis there is a corresponding number. This property is easy to visualize whether one axis or two are used. All the points are "filled up" with numbers —irrational, rational, positive, and negative. For each number there is a corresponding point, and vice versa. This correspondence between points and numbers is one of the basic premises of coordinate geometry—although Descartes' analytic mind failed to realize it.

The foregoing seems perfectly innocent. Now comes the difficulty. The points and numbers on the line can be raised or lowered so that for every point in the quadrants or spaces above and below the axis there is a corresponding point on x (Diagram 9). Now consider the line OP, which is not parallel to the x axis (Diagram 10). The line ends at the point which corresponds to

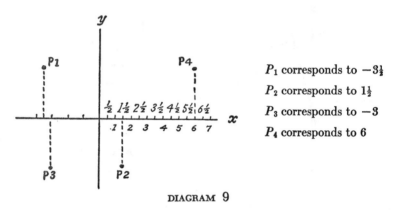

P_1 corresponds to $-3\frac{1}{2}$

P_2 corresponds to $1\frac{1}{2}$

P_3 corresponds to -3

P_4 corresponds to 6

DIAGRAM 9

the number 4 on the x axis. According to our point-number correspondence, there should be a number for every point on OP, and this number is determined by the corresponding number on the x axis. Yet it is obvious that OP is longer than the part of the x axis that determines it. How can there be enough points (or numbers) on the axis to fill up all the points on the longer line? The number field on x apparently is too small. Yet logic tells us that by raising or lowering the x axis, we can account for every point in space above or below.

Fortunately Descartes never saw this paradox, and by the time it was noticed, analytic geometry had proved itself so worthwhile that mathematicians could not scrap it. Instead they scrapped the paradox, ignored again the old problem of infinity creeping out on the stage in a different costume. For an answer to the puzzle mathematicians had to wait two centuries; the present reader has only to wait a little more than a hundred pages. Meanwhile, he can ruminate on a clue: Which are there more of, natural integers or even numbers?

Two other important mathematical achievements of Descartes might be mentioned, although they are minor compared to his creation of analytic geometry. The first is the formulation of rules

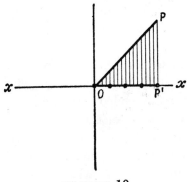

DIAGRAM 10

Are there more points, i.e., numbers, from *O* to *P* than from *O* to *P'*?

for transposing numbers from one side of the equals sign to the other and of determining how many roots of an equation are positive and how many negative. Actually Descartes lifted this discovery from an English mathematician, Harriot. As was the case with Cardano, the plagiarizer was credited by having the rule named after himself. Descartes' Rule of Signs is familiar to every algebra student.

The second achievement was Descartes' method of labeling constants (coefficients) and variables (unknowns), the former by the first letters of the alphabet and the latter by the last letters. For instance, the equation $3x^2 - 4x + 1 = 0$ can be written in the more general form as $ax^2 + bx + c = 0$, where a, b, and c represent the coefficients 3, -4, and 1, respectively, and x represents the unknown. This literal notation, which had been the unconscious goal of mathematicians for centuries, changed algebra from the study of particular equations to the study of general ones. Mathematicians could hereafter classify equations according to types instead of dealing with a vast hodgepodge of individual problems. All linear equations, for example, fall into the general form $ax + b = 0$; quadratic equations into the form $ax^2 + bx + c = 0$; cubic equations, $ax^3 + bx^2 + cx + d = 0$; and so forth. Furthermore, any equation—linear, quadratic, etc.—can be rendered in the most general form: $ax^n + bx^{n-1} + cx^{n-2} + dx^{n-3} + \cdots + px + q = 0$.

A completely literal notation represents the final step in the evolution of the language of mathematics. What had begun with problems stated in words, ended with problems stated in symbols. Algebra had become, as Antoine Lavoisier wrote, "the analytical method *par excellence*: it has been contrived to facilitate the operations of the understanding, to make reasoning more concise, and to contract into a few lines what would have needed many pages of discussion; in short, to lead by a more convenient, rapid and unerring method, to the solution of the most complicated questions."

The evolution of a literal notation led to a generalization and systematization of equations. It also enlarged the field of equations that could be studied by obliterating the artificial restrictions that had been set up. These restrictions against the use of negative and imaginary numbers have been noted previously. Mathematicians had not imposed these restrictions without reason. Quite the contrary; the restrictions were necessary when problems were stated in words. For instance, if a farmer has 10 cows and a certain number run away and he slaughters three-quarters of the remaining cows, leaving him with 5 cows, how many ran away? The answer is that $3\frac{1}{3}$ cows ran away. Obviously, this is impossible. No one has $3\frac{1}{3}$ cows or a negative, irrational, or imaginary number of cows. Therefore, early algebraists excluded equations that yielded these impossible answers. Late into the Middle Ages, mathematicians still did not consider anything but an integer as a number and only problems that could be answered by a positive integer were considered soluble.

When equations are written completely in literal notation, however, their impossibility is less evident, for just as one cannot tell who are the "impossible" people in a list of unfamiliar names, so one cannot distinguish the "impossible" equations when their identity is hidden behind letters of the alphabet. Unwittingly, mathematicians began admitting these "impossible" equations and their relatives, negative numbers, irrationals, etc., and found them to be just as desirable as the more familiar integers. Thus, by symbolizing what was "impossible," literal notation helped integrate a whole body of outcast equations and numbers into mathematics. All the prejudice against them had been unwarranted.

Descartes' coordinate geometry was released to the world in 1637, almost twenty years after he had begun it. Its debut could not have been more modest; the geometry was the wallflower of his ideas. He appended it to what he considered the more weighty *Discourse on Method,* an essay interlarded with biographical detail and describing his scientific approach. The treatise *Geometry* was tacked on at the end along with two other essays to illustrate what sort of results the "method" could achieve. If Descartes had realized the import of his geometric Cinderella, he would undoubtedly have decked her out in better finery and given her her own coach.

Copies of the *Discourse,* with the three essays in the appendix, were sent to most of the prominent mathematicians. Immediately Fermat, another Frenchman, who was certainly Descartes' mathematical equal—if not his superior—objected to trivial points. He was supported in these objections by his friend Roberval, who may have been overly critical because he had been slighted by not receiving a complimentary copy. Blaise Pascal, another seven-

teenth-century mathematical genius, and his father, Étienne, condemned the whole geometry and refused to admit its value. Their
objections, too, may have been more personal than academic, for
Descartes had once publicly declared that Pascal *père* had written a treatise on conic sections and passed it off as the work of
sixteen-year-old Pascal *fils*. All the petty intrigues, personal feuds,
and backbiting of the inbred French mathematicians could not,
however, keep the new geometry from the audience awaiting it.
Within a few years it was being taught in almost every university
and no man who called himself a mathematician could ignore it.

With the publication of the *Discourse on Method*, Descartes'
fame spread. Like the author, most readers considered the *Discourse* rather than the essay on geometry as more important. In
Holland Descartes' reputation was at such a height that one of his
disciples managed to convert almost the entire faculty of the University of Utrecht to the Cartesian philosophy. One of the professors was so awed by the power of Descartes' ideas that he
begged the great philosopher to let him take the place vacated by
a friend who had recently died, declaring that it would put him
in "the Third Heavens." A Dr. Henry More, "whose Passion and
Reverence for our Philosopher had almost proceeded to Idolatry,"
became a disciple and missionary for the Cartesian philosophy in
England. Royalty began to woo him: Charles I of England and
Louis XIII of France invited him to adorn their respective courts.
Elizabeth, Princess of Bohemia, became a dedicated correspond-
ent and pupil of his.

Unfortunately, the other side of the coin of adulation is jealousy,
and Descartes reaped his share from a man "of an ordinary mean
Judgement and Superficial Learning" who went about with a
"triumphant huffing Air." Here was a man, Voetius by name,
whom Descartes' contemporary biographer called "maggotish
. . . and prone to contention and wrangling." Despite these un-
Christian shortcomings, Voetius had managed to become head of
the Divinity department at the University of Utrecht. Voetius

wrote several papers in which he described all the belief of atheists and ingeniously, "scattered up and down in his Theseses amongst the *Criterians of Atheism,* all the things he knew to be ascribed to M. Des Cartes."

These charges sound extraordinary in the light of Descartes' obsequious submission to the Jesuits. Yet Voetius was one of the few men who realized where the Cartesian philosophy would lead: to the destruction of religion. His antagonism toward Descartes seemed to be promoted more by genuine difference of opinion than envy, despite what Descartes' biographer said. It is ironical that a man as devout as Descartes should have unwittingly erected a philosophy that purports to prove the existence of God and that in the long run should destroy Him. But that is just what he did. By appealing to reason to determine what is truth, Descartes denied the ability of the Scriptures and the Church to determine it, and although he managed to "prove" the existence of God, it did not take long for this proof to crack at its prefabricated seams. Theologians soon realized that the scientific method cannot be used to prove God—if anything, it casts Him out of the picture entirely by giving scientific explanations for what were once considered acts of God.

Not only does the Cartesian philosophy oust God from His heaven, it banishes all nonrational truth to limbo. Painting, music, poetry, literature—all are of no importance in Descartes' scheme. They are incapable of teaching us anything. Even the lessons of history are unimportant next to what can be learned by using the rational method. Reason, not experience, is what counts.

Superficially, Descartes' works are pious. Essentially, they are heretical. During his lifetime he was, for the most part, hailed by the Church; yet a few years after his death he was condemned and his books placed on the Index. The effect of Cartesianism was not only heretical, but strangely enough, puritanical as well. For the next century or two, almost all art bowed to reason and attempted to ingratiate itself by instructing rather than by giving pleasure. A clear, bare style without flourishes became the ideal. Novels were preferred to poetry, for a novel could teach while poetry, according to John Locke, could only depict "pleasant pictures and agreeable visions." Not until the late eighteenth century did men challenge the sterile Reason that Descartes had enthroned. Wordsworth and Blake, for example, were among the first to react against the "intellectual all-in-all" who "would peep and botanize upon his mother's grave."

Since Descartes had shut emotion out of his life, it is little wonder that his philosophy does the same. Even the quarrel with Voetius failed to arouse him, although his adherents were infuriated at "such a dirty, smooty Behaviour" and took up the cudgels. The controversy, which only added to Descartes' fame, continued for ten years, with new recruits being added to both sides. At one point Descartes' adherents were forbidden to teach the newfangled notion of circulation of the blood—this being one of Descartes' favorite discoveries. Despite the "needless Cavils and gross calumniating Abuses" against him, the great mathematician remained surprisingly neutral, preferring to "bury all the whole Affair in Oblivion." Only when the Jesuits might be swayed by his adversaries did he take any steps to protect himself.

It was while he was involved in the controversy with Voetius that a secret tragedy struck. His five-year-old daughter, Francina, died. As a rule, Descartes was more of a thinking machine than a man. Learning and knowledge were his passions; emotion had no place in his philosophy or in his life. But when his daughter died, he was overcome and realized, perhaps for the first time, that "Philosophy cannot stifle Natural Affection."

A record of Francina's life and death was found written in the flyleaf of one of his books. It is the only way we know that she existed at all, for Descartes kept his personal life so private that even the full name of his child's mother is unknown. There is no record to indicate that he ever married or had any other children

—although he was sometimes given credit for having had more: "He was too wise a man to encomber himselfe with a wife; but as he was a man, he had the desires and appetites of a man; he therefore kept a good conditioned hansome woman that he liked, and by whome he had some children (I think 1 or 2)," wrote the gossipy biographer, John Aubrey.

After Francina's death, Descartes was offered a pension by the King of France, with the stipulation that he return to his native country to live. Descartes accepted, but when he presented himself at the French court, everyone just stared "him in the face, as if he had been some Elephant or Leopard" and no one knew anything about the pension. So back he went "to his dear Egmond" in Holland, determined not to trust "bare promises" of emoluments in the future.

When Queen Christina of Sweden pleaded with him a few years later to come to her court, he refused. But the Queen was not put off easily. Her penchant for surrounding herself with scholars had reached the point where it "began to become the object of scorn, railery and obloquy of Strangers; it was in every bodies mouth that she assembled all the Pedants of Europe to Stockholm, and that it would not be long ere the government of the Realm would be managed by Grammarians." And now she wanted the most famous of them all—Descartes—to add to her collection. Enticed by promises and praise, Descartes finally relented —not without misgivings—to go to "the land of rocks and bears" for three months.

In the fall of 1649 he set out for Sweden and on his arrival was received by the Queen "with such a distinction as was observed by the whole Court." Then for six weeks Christina ignored him— supposedly so that he could leisurely "familiarize himself with the genius of the Country."

Although Descartes had planned to stay only three months, the Queen had other ideas. She appointed him to her Privy Council— obviously an action with long-range implications—and commanded him to "compile a compleat body of all his Philosophy," also a task that would keep him busy for more than a few months. Descartes accordingly "began to rummish and ransack his Trunk" of notes and writings, but his papers "were all in pieces scattered up and down," and he never got to put them in order.

Christina had great plans for her newest and most glittering catch. She planned to give him a house in the most southerly and warmest part of Sweden, present him with a title, and best of all, build an academy and appoint him director. Certainly, with all these inducements, it would be difficult for him to leave.

After his initial six weeks of freedom, Descartes was summoned daily to instruct and converse with the Queen. Unfortunately, the sessions began at five in the morning and were held in a large, cold room—not too pleasant an arrangement for a man who was used to staying in bed until almost noon and who hated the cold. But Descartes was either too timid, too polite, or too awed by royalty to complain. After two and a half months of the new regime, he fell ill with "Feaver, accompanied with an Inflammation of his Lungs"—probably pneumonia. For a week he lay delirious and close to death, and then, on "the 11th of February at Four a Clock in the Morning, Aged 53 Years, 10 Months, and 11 Days," he became conscious long enough to speak what are purported to be his last words: "My soul, thou hast long been held captive; the hour has come for thee to quit thy prison, to leave the trammels of this body; suffer, then, this separation with joy and courage." With that splendid speech, he died.

He was buried in Stockholm, where Queen Christina was still going ahead with her grandiose plans for the great philosopher—albeit somewhat altered. She wanted to build him a grand mausoleum, but the French ambassador dissuaded her. Descartes' body was later moved to Paris, where it rested for a while before being moved twice more.

Blaise Pascal
1623-1662

A new world had been born: a Renaissance world, a world of science, a world that came out of the womb of medieval darkness and into the light of the modern age. And like all births, this one was accompanied by pain. Savonarola, an Italian ascetic, had been burned in a public square in Florence because he refused to help in the birth of this new world. Bruno, a hundred years later, had been burned at the stake and Galileo thrown into prison because they would help. All over Europe Inquisitions had been set up to try to keep the old order. And all over Europe schools and universities and new religious groups had been set up to effect the change. Europe had been reformed. Princes and kings of the land had wrenched power from the princes of the Church, who now tried to wrench it back again. The Counter-Reformation came into full bloom and those who tried to serve both the princes and the popes, the kings and the cardinals, found themselves in trouble. In England Henry VIII made himself sovereign of both church and state, thus solving the problem of allegiance for his subjects. In France it was still to be solved.

It was into these two conflicting worlds—the spiritual and the secular—that Blaise Pascal was born in 1623, on June 19, in Clermont, France. His whole life was to be a constant struggle between one world and the other. An egotistical, conceited fop, he grasped for fame and fortune in the world of the flesh. A pious, almost fanatical kind of saint, he strove for grace and salvation in the world of the spirit. Eight years before he died the battle ended.

The Pascals came from central France where Étienne, Blaise's father, an important government official, was a member of the new nobility of France, a class noted for its ambition, worldliness, and intellectual attainments. He and his wife had three children, Gilberte, the oldest; Blaise; and Jacqueline, the youngest.

As a child Blaise was sickly, supposedly due to a witch's spell. His niece Marguerite related that at the age of one he suddenly became very weak and would scream hysterically at the sight of water or at seeing his parents embrace. This so-called "death spell" that had been put on him was transferred to a cat by the obliging witch and little Blaise was spared. Marguerite went on to say that the cat fell out of a window and was, in accordance with the spell, killed on the spot. If anything, the story illustrates the superstitious thinking that pervaded the early seventeenth century.

When Blaise was three, his mother died, and shortly after, Étienne shook the dirt of Clermont from his heels and moved what was left of his family to Paris where he started his children on their unusually progressive educations. None of the Pascal children ever went to school. Instead, they were taught at home by their father and an occasional tutor. *Père* Pascal had his own ideas about education, namely that children should not be pushed—they should never study a subject until they could master it easily. He felt that natural curiosity—not a stern pedagogue—should lead students along the road of learning. In accordance with these ideas, the Pascal children did not start languages until they were twelve, and mathematics was supposed to be a dark secret until they were sixteen. But Blaise's natural curiosity was more than his father anticipated. Forbidden to study geometry, he became so intrigued that he decided to investigate the subject by himself, drawing diagrams and figures on the tiles of his playroom floor, and inventing his own terminology—"rounds" for circles and "bars" for straight lines. Working by himself "he pushed his researches so far that he reached the thirty-second theorem of the first book of Euclid" before his astonished father discovered what he was doing. Astounded and pleased, Étienne gladly presented his son with Euclid's *Elements* and proudly presented his friends with his little genius.

Blaise was not the only prodigy in the family. Jacqueline, with her pretty face, winning ways, and unusual flair for writing poetry, came in for her share of the limelight. Through friends her poems were brought to the attention of the royal court where the queen was so captivated by the lovely, talented child that she invited her to the palace time and time again. Both Blaise and Jacqueline relished their extraordinary positions. Worldly acclaim was sweet to them.

When Blaise was fifteen his father got into trouble with the government over some payments due him. Several of the other "conspirators" were thrown in jail and to escape the same fate, Étienne Pascal fled, leaving his children alone in Paris. No sooner was he gone than Jacqueline came down with smallpox and Papa Pascal rushed back long enough to nurse her and then sneaked off to Clermont. Smallpox left Jacqueline's pretty face pitted with scars—a not uncommon thing in those days. With more optimism than seems warranted, she wrote a poem to God thanking Him for protecting her innocence by scarring her face.

Despite the disfigurement, her popularity at the palace was undimmed and she was called back to act in a play. She cleverly used the opportunity to regain favor for her father. After the entertainment, Cardinal Richelieu—the man from whom Étienne was hiding—took Jacqueline on his lap and congratulated her on her superb acting ability. Being a better actress than he realized, Jacqueline immediately burst into tears. When the distressed Cardinal discovered that she was crying because she missed her father, he dropped the charges against Étienne, allowed him to return to Paris, and shortly after had him appointed tax commissioner for the Normandy region.

Blaise was sixteen at the time and had just issued his first important writing—a pamphlet on conic sections that had caused quite a flurry in mathematical circles. The great Descartes was so impressed that he could not believe that a mere teenager had developed the subject further than anyone since Apollonius. The name of Pascal was on every tongue in every salon where mathematics was discussed.

In connection with his work on conic sections, Pascal studied projective geometry which had been recently invented by an architect-engineer, Gérard Desargues. An eccentric who bragged that he had never read a book in his life, Desargues bequeathed his own poor reputation on his brain-child. Shunning conventional terminology, he named his figures after trees and flowers with the result that projective geometry was ignored or ridiculed by his contemporaries. Pascal seems to have been the only man to realize that the subject was legitimate and not the freakish offspring of a half-cracked charlatan.

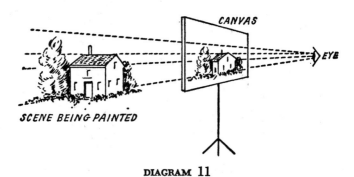

DIAGRAM 11

Projective geometry had been practiced for at least a century before Desargues and Pascal converted it into a formal study. Artists had used the idea of projection in order to paint pictures

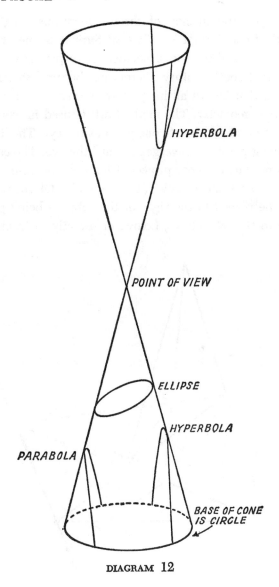

HYPERBOLA

POINT OF VIEW

ELLIPSE

HYPERBOLA

PARABOLA

BASE OF CONE
IS CIRCLE

DIAGRAM 12

that conveyed the illusion of three dimensions—height, width, and depth. Leonardo da Vinci's *Last Supper* is one of the best-known examples of the use of perspective or projective geometry. Uccello, a fifteenth-century Florentine, became so adept with perspective that almost all his pictures are simply tours de force in projective geometry. The method artists used in working with perspective was to designate one point as the eye. The lines of the objects being painted converged toward the eye. The canvas was a section or slice of these projected lines cut off somewhere between the object and the eye (Diagram 11). Of course, the eye need not be placed to the right of the subjects being painted, it could be to the left, above, below, or directly in front of them.

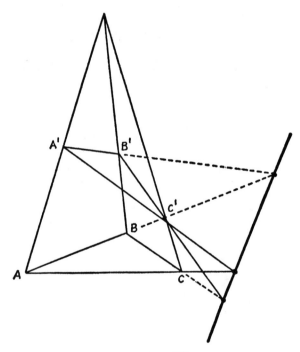

DIAGRAM 13

Desargues and Pascal began studying projections as a new kind of geometry instead of as an artistic device. They found that all conic sections—ellipses, hyperbolas, parabolas—were projections of a circle viewed from the point of a cone (Diagram 12). They investigated what properties projections have in common with their various sections. For instance, in Diagram 13, what do triangles *ABC* and *A'B'C'* have in common? Desargues proved that the corresponding sides all meet in three points that lie on one straight line (Diagram 13).

One of the most interesting theorems proved by Pascal concerns what he called "the mystic hexagon." If a hexagon is inscribed in a circle and the sides are extended (Diagram 14), then *AF* and

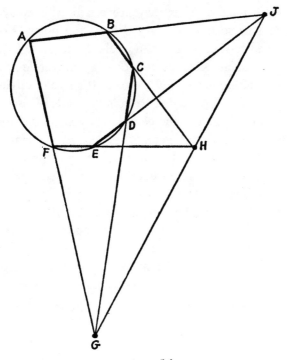

DIAGRAM 14

CD, BC and *FE, ED* and *AB* meet in three separate points *G,*
H, and *J,* which all lie on the same straight line. If a section is
made of the circle and hexagon, then the sides of this new hexagon, *A'B'C'D'E'F',* will also meet in three points that lie on a
straight line (Diagram 15). Furthermore, this second line *G'H'J'*
is a projection of the first line *GHJ.* By using his "mystic hexagon,"
Pascal claimed that he could derive over four hundred corollaries!

Is projective geometry an independent and ingenious new
geometry, or is it part of Euclidean geometry? Where does it "fit
in" in mathematics? Such questions are valid and important.
Descartes' analytic geometry showed that even the most different
branches of mathematics—geometry and algebra—connect. Certainly it is not too much to assume that projective and Euclidean
geometry are related. That they are can be shown by taking a
circle *A* and its section *A',* viewed from the point *O* of a cone.
If *O* is moved farther away, then circle *A'* increases in size while
A remains the same. Keep moving *O* away and circle *A'* continues
to increase, getting closer and closer to the size of circle *A* (Diagram 16). When *O* is moved to infinity, circle *A'* is exactly the
same size as circle *A* and the cone has become a cylinder. When
O is at infinity, projective geometry becomes the same as Euclidean geometry—thus, Euclidean geometry and projective geometry are related to each other in infinity. Euclidean geometry does
not contain projective geometry—it is a branch of it. More recently
it has been discovered that projective geometry is not the whole
but is itself only part of an even larger study, topology.

Projective geometry was ignored by everyone except Pascal
and Desargues. Mathematicians were more interested in Descartes' new analytic geometry. But today, projective geometry
has come into its own and serves a very useful function in mathematics. For instance, projective geometry investigates what a
circle has in common with one of its sections, an ellipse. Through
projection, the circle has been *transformed* into an ellipse, yet
certain things remain *invariant.* In the modern theory of rela-

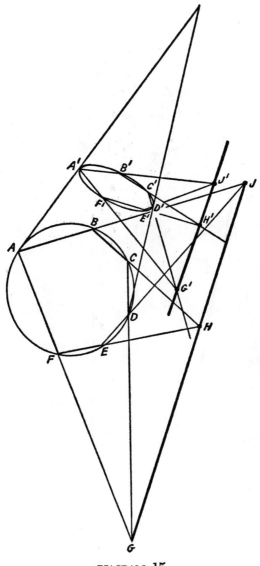

DIAGRAM 15

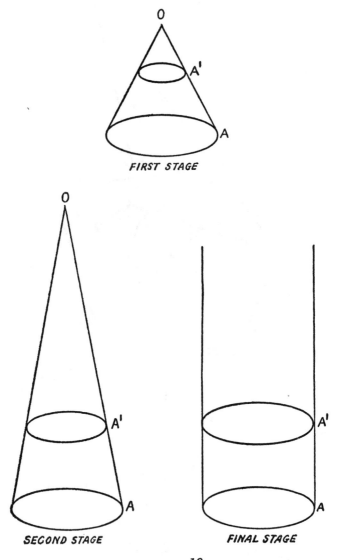

DIAGRAM 16

tivity, different bodies have different time-space worlds. The geometry of one world is different from that of another, yet physicists want to know what laws are *invariant* in these different worlds. Projective geometry shows them how to go about finding these invariants.

Through analytic geometry and through transformations and invariants, projective geometry is related to algebra. Algebraic equations can also be transformed, just as circles can be transformed to ellipses, but something—an invariant—remains the same.

After the publication of Blaise Pascal's treatise on conic sections, the Pascals settled in Rouen, where Étienne was to take up his position as tax collector. Blaise was still flushed with success. He was not only on speaking terms with all of France's greatest mathematicians, Descartes, Fermat, and Roberval, but was of equal standing with them. And he was only sixteen years old.

The sophisticated, witty, conceited boy was unimpressed by the riots and ensuing executions that greeted his father in Rouen. He was too absorbed in his work to notice anything except possibly the adulation that was being showered on his sister. Jacqueline had continued with her poetry and entered a local competition which she won—the first girl ever to have done so. The judges marveled that a mere child of twelve could write so well. Actually she was fifteen—the Pascals apparently were not averse to subtracting a year or two from their ages to make their precocious accomplishments seem even more so. Blaise complained that Jacqueline should not be seeking "the love of the world" while he was shut up helping his father add the long dreary columns of figures that were part of the tax commissioner's job. Nevertheless, Blaise himself had a scheme that he hoped would not only ease his tedious work but would bring him fame and fortune. He abandoned his geometry to work on this new, more practical idea: a calculating machine.

By the time he was eighteen he had perfected it to the point where it could add, subtract, multiply, and divide, and even planned to have it extract square roots. This machine—the first of its kind—caused a sensation. News of it quickly spread and Pascal was called on to demonstrate it to various interested nobles and mathematical groups. He patented the machine—after some feuding with a local clock-maker who tried to steal the idea—and hoped to make his fortune selling it. But the exorbitant cost of manufacture made the venture unprofitable.

Life in Rouen was undoubtedly routine and dull after the glitter of Paris and the stimulation of the weekly mathematical meetings the Pascals had attended there. Blaise continued working and studying and then, when he was twenty-three, a minor event occurred which was to have a major effect on his life and the lives of the whole family for two generations. Étienne slipped on a patch of ice and hurt his hip. Two bone-setters were called in from a nearby hospital and these two men, brothers, stayed on at the Pascal house for three months, ostensibly to nurse Pascal *père*. But more than bone-setters or nurses, they were converts to a fanatical religious sect called Jansenism, which had been founded a few years before by a Flemish man named Jansen. Absolute and extreme in their uncompromising attitude toward religion, they preached that strictness alone could save the Church and its thousands of lost souls. Man must strive for ever higher states of Grace; to stand still is to slip back. (An echo of this idea is found today in the preachings of corporation directors to their flock of stockholders: "A business must continue to grow or it will slip back swiftly. There is no standing still.") And so it was with the business of soul-saving, where the coin of exchange is faith. Faith alone revealed God to man, despite Descartes' assertions to the contrary. Reason had no place in the doctrine of the Jansenists.

Incredible as it may seem, the two bone-setting fanatics managed to win Blaise Pascal as a convert. Pascal—a man whose whole outlook was scientific and rational; a man who trusted reason before anything else; a cynical, egotistical, fame-craving man wholeheartedly embraced this austere religion that was the antithesis of everything he had been taught.

Why? Perhaps the best answer is in his own words: "The heart has reasons that the reason knows not of." Religion is an affair of the heart, the spirit; science, an affair of the mind. The two were to him distinct, separate, not joined together in the scholasticism of the past. Therefore, a man could believe in both religion and science without contradicting himself.

Descartes had tried to join religion and science. The result was not successful. If religion were to survive it would have to do so by standing on its own feet and not by grafting itself onto the body of science. Pascal, unlike many of his contemporaries and successors, realized this.

Although he had always been a Catholic, Blaise had been neither especially devout nor attracted by the growing band of atheists and agnostics. Yet at the age of twenty-three, under the proddings of the two bone-setting Jansenists, he gave up his indifference and became fanatically religious. Science was no longer important to him—only religion mattered. Here again we may ask "Why?" and answer it in his own words: "When I ponder on the short span of my life, swallowed up in an eternity of the past and the future, the trifle of space that I fill and the only one I can behold, sunk in the infinite reaches of space, which I know not and know not me, I am terrified with wonder to see myself here instead of somewhere else . . . Who has placed me here? By whose order and warrant was this place and this time ordained for me? The eternal silence of these infinite spaces leaves me in terror."

These are questions that science cannot answer and are greater than any science even asks. Intellectual exercises cannot develop man's spiritual muscles. What is food for thought cannot satisfy the hungers of the heart—and Pascal was emotionally starved.

He was the neurotic type of person that must go "all the way."
A compromising, watered-down religion that would be easy to
follow was not for him. If he had to dedicate himself to something,
it would be with all his being, not just his left hand. Everything
he ever did, he did with a frenzy, becoming completely consumed
in the project at hand, whether it was mathematics, physics, religion, or a social life.

That Blaise's sisters—as well as he—were nothing but chips off
the old block seems to be indicated by the fact that in short order
he succeeded in converting Jacqueline and then the two of them
recruited their father. Gilberte, who had since married and moved
to Clermont, came to Rouen for a visit with her husband, and
they too were won over to the Jansenist camp.

Not long after his conversion, Blaise took a leading part in the
persecution of a local friar whose ideas differed from those of the
Jansenists. Mercilessly, with all the fanaticism of a fresh convert,
he led the fight to have the friar recant his heresies. And Blaise
was no easy man to have as an enemy. His intelligence and supreme overconfidence in his own superb abilities were enough to
frighten anyone. "Those who did not know him," his sister wrote,
"were at first surprised when they heard him in conversation
because he always seemed to take charge with a sort of domina-
tion." The poor friar risked being burned alive if he were proved
guilty of Pascal's charges. The affair fortunately ended without
any loss of life.

After a few months Blaise's religious zeal waned and he resumed his scientific experiments again. He had learned of the
revival of interest in vacuums—could they exist or could they not
exist—and immediately threw himself into the fray by trying
to figure out a way to prove that a vacuum can exist. Aristotle
had claimed that "Nature abhors a vacuum" and therefore a
vacuum cannot exist. The more experimental scientists, like Pascal, pooh-poohed this classical nonsense and declared that vacuums can exist.

Pascal's experiments came to a temporary halt when his health began to fail. Ever since the age of eighteen he had been sickly, but now he was confined to bed, paralyzed from the waist down, barely able to swallow enough food to keep himself alive, and tortured by severe headaches.

That summer, just as he turned twenty-four, he and Jacqueline went to Paris where the medical care was supposedly better. Still confined to his bed, Pascal was visited by many an eminent scientist who had heard of his vacuum experiments and his ingenious calculating machine. The great Descartes came and sat at the side of his bed, gave him some medical advice, and then discussed the vacuum theory that was currently rippling scientific circles.

By October Blaise was well enough to go on with his work. To test the theory of a vacuum, a long tube—at least thirty-one inches —was filled with mercury and then, with a finger held over the open end, the tube was inverted and placed into a pan also filled with mercury. When the finger was removed from the opening, a little of the liquid from the tube flowed into the pan, leaving a space at the top of the tube. The problem was then to ascertain what was in the space: air? mercury fumes? nothing? or "subtle matter"? as Descartes claimed. When the tube was raised and lowered in the pan of mercury, the mercury in the tube remained at the same level, giving rise to the hypothesis that the pressure or weight of the air on the mercury in the pan equaled the pressure of the mercury in the tube. But even the idea of the existence of air pressure was unproved.

Pascal continued to conduct very elaborate experiments with tubes inside tubes, tubes filled with different liquids, and tubes of varying shapes and sizes—some as long as forty feet. Finally he hit on the ingenious idea of measuring the height of the mercury in a tube at the top and bottom of a mountain. If there were a difference in the height of the mercury, it could be attributed to air pressure, for it would be ridiculous to say that any differ-

ence was because "Nature abhors the void more at the foot of
the mountain than at its top." It would be more logical to assume
that the pressure of air is greater at a lower altitude than at a
higher and thus causes a smaller or larger vacuum. The test was
carried out, and as the altitude fell, the mercury rose in the tube,
proving that air does have weight and that this weight can lift
or lower the mercury inside a tube that is placed, open end down,
in a pan of mercury. The corollary is that as the mercury drops
in a tube, it creates a vacuum. Pascal published his results and
was immediately charged with stealing the whole idea of the
experiment from Descartes.

While Pascal was conducting his experiments, Jacqueline was
becoming more and more entranced with Jansenist teachings, and
by 1648 had decided to become a nun. Her brother concurred in
the idea, but when *Père* Pascal arrived in Paris, he would hear
nothing about the scheme and grumbled that Blaise was responsible for putting these crazy ideas into his sister's head. Nevertheless, Jacqueline was determined to lead a religious life, even
if her father refused to let her enter the convent. She lived at
home like a nun, staying in her room alone and praying all day.
Realizing that her attack of religion was both acute and chronic,
Étienne made her promise not to join an order until after he died,
"for his life would not be very long now."

The following year the Pascals returned to their home town of
Clermont for a visit. Jacqueline, as usual, devoted herself to religion. She turned her literary talents to the same end by translating
Latin prayers into French for a local priest, who was so pleased
with the results that he urged her to continue. But when Mother
Agnes at Port-Royal, the Jansenist center near Paris, heard about
it, she wrote saying, "That is a talent of which God will ask you
no accounting; you must bury it." Jacqueline obeyed.

Blaise, some months earlier, had been ordered by his doctors
to avoid work and to "seek diversion in society," an order which
he likewise obeyed to the hilt. Parties, gambling, and amusements

filled his days. His wit and gaiety attracted many friends, and the great fabulist La Fontaine wrote that "the brilliance of M. Pascal charmed and enchanted everyone." At Clermont he supposedly squired the "Sappho of this region" about. When he returned to Paris, he took up with several high-living friends, among them the Duc de Roannez, who introduced Blaise to "the delicious commerce of the world." This delicious commerce was not as licentious as we might image. Pascal indulged himself with fine clothes, horses and carriage, good food and wines, gambling, and possibly drinking, but never women. Even without supporting a mistress, however, his standard of living was higher than his income, and his father held a tight grip on the purse strings.

Pascal needed money, and to get it he once more tried to make a profit from his calculating machine by interesting Queen Christina of Sweden, to no avail. Then in September of 1651, his problems were temporarily solved by the death of his father. Blaise was still enough of a religious know-it-all to write to Gilberte, "Let us not grieve like the pagans who have no hope," and proceeded to set a good example by stepping up his fast living to an even faster pace.

Jacqueline was now free to enter the convent at Port-Royal, which she did promptly, but not before unwittingly agreeing to let Blaise handle her inheritance. In less than two years she was ready for her final vows and wrote her brother, asking that he return "the little property which God has given me" so that she could use it as her dowry. Blaise, who had been the original instigator of Jacqueline's religious fervor, now opposed her becoming a nun and refused to give her her share of the inheritance. This meant that she would have to become a "lay sister," working as a kitchen maid or scrub woman instead of a "bride of Christ," devoting her time to prayer. Blaise's rascality finally lost out to family pride and he gave Jacqueline her money. The day after her irrevocable vows were made, he wrote his brother-in-law, saying, "My sister made her profession yesterday. It was impossible for me to prevent it."

The next two years for Pascal were carefree ones which Gilberte dismissed as "the time of his life that was worst employed." It was also in these two years that Pascal collaborated with Fermat to develop the theory of probability.

Like Cardano, Pascal became interested in probability through gambling. A dice game gave rise to the questions of what the chances of a certain throw on a pair of die were and of how the stakes should be divided if the game were stopped before it were finished. Pascal took the matter up with his father's old friend, an amateur mathematician, Pierre Fermat. Together, the two of them laid the foundation for a whole new area of mathematics that is of great importance today in a vast number of fields: calculating insurance risks, studying the behavior of gases, interpreting statistics, studying heredity, to name only a few.

Probability is "simply the general problem of logic," as Charles Peirce, an eminent American mathematician, remarked. It consists first of determining in how many different ways an event can occur. The probability of one of these ways' occurring is the ratio of the number of these specific ways to the total number of ways. Thus, if one wants to determine what is the probability of a coin landing heads up, one must first determine the total number of ways in which a coin can fall. It can fall heads or tails, i.e., two ways. (We shall overlook the remote possibility that it might land on its edge.) The probability that the coin will land heads up represents one of the two ways. Therefore the probability of getting a head on a throw is 1 in 2, or $\frac{1}{2}$. The probability that the coin will land either heads or tails up is 2 in 2, or 1. Thus, if an event is sure to occur, its probability is 1. As the probability decreases, the ratio decreases and finally, if there is no possibility of an event's occurring, its probability becomes zero.

Pascal concocted a "magic triangle" which tells at a glance the probability of an event's occurring. The triangle looks like this:

$$1$$
$$1 \quad 1$$
$$1 \quad 2 \quad 1$$
$$1 \quad 3 \quad 3 \quad 1$$
$$1 \quad 4 \quad 6 \quad 4 \quad 1$$
$$1 \quad 5 \quad 10 \quad 10 \quad 5 \quad 1$$

The top row has only the number 1 and represents the case where an event cannot fail to happen. All the outside numbers in the other rows are 1 and the rest of the numbers are filled in by taking the sum of the two numbers just above each one; thus, in the third row, the sum of the two numbers above the center is $1 + 1 = 2$.

This triangle is extremely useful in determining simple probability problems such as those dealing with coin tossing. More complex problems would need a triangle with hundreds of rows, but other methods than magic triangles have been devised to find probability ratios. To determine from Pascal's triangle the probability of two heads appearing on two tossed coins, we look at the third row (if three coins were being used we should look at the fourth row; four coins, the fifth row; and so on). The sum of the numbers in the third row is 4, which gives us the total number of ways in which the coins can fall: two heads, two tails, a head and a tail, and a tail and a head. The chance of throwing two heads is 1 in 4, or, the first number divided by the sum of the numbers. The chance of one head is 2 in 4, or the second number divided by the sum; and the chance of no heads is 1 in 4, or the third number divided by the sum.

It is important to note that of the four possible ways in which the coins can fall, two are alike—or seemingly alike—a head and a tail, and a tail and a head. In both cases we get only one head, yet they are on different coins, and must be counted as different results. In figuring all probability problems of this type it is important to include *every possible way* in which the throw or event can take place before figuring the odds for the occurrence of a particular throw or event. Furthermore, each event must have an equally possible chance of occurring. For instance, if a lottery ticket is bought, it can be either the winning or losing ticket, but this does not give the buyer a 50-50 chance of winning. The chances of holding a losing ticket are much greater than of holding a winning ticket—common sense tells us that. According to the laws of probability, *all* possibilities must be included—that is, if 10,000 tickets are sold and one is the winning ticket, there are 10,000 different tickets or ways in which the event can happen. The particular event of holding the winning ticket occurs only once in these 10,000 ways.

The influence of the laws of probability on modern thinking is far greater than can be indicated by simply discussing the throwing of dice or sale of lottery tickets. All modern life is based on the premise that these laws will continue to operate. They are part of our belief in an ordered universe. A famous short story by Robert Coates describes what happens when the law of averages ceases to function.* Suddenly the highways are crowded with everyone who owns a car, all deciding to take a pleasure drive on exactly the same night. The next day there is a sudden run on the stores for spools of pink thread—every woman who sews suddenly decides she needs this particular color.

An even more fantastic example of what would happen if these laws failed to operate might be found in the motion of molecules. Every object is made up of molecules that are constantly moving in different directions, some whirling off away from the "body" of the object. Now, what would happen if *all* the molecules, contrary to the laws of chance, suddenly whirled away? The object would literally disappear before our eyes. The probability that such a thing could happen does exist but is so infinitesimal that we do not even consider it.

* Coates, Robert, "The Law," *The New Yorker*, November 29, 1947.

It was this far-reaching work in mathematics that Pascal was doing during his profligate years—the years in which he was "worst employed." Both pleasure and science were the enemies of Jansenism.

During this so-called "worldly period" Pascal courted an unknown young lady, but society palled on him before anything came of it. Two of his friends noted that he was on the brink of "a very advantageous marriage" when his life underwent a drastic change. Toward the end of 1654, Pascal "was seized by a great contempt for the world and an unbearable disgust for all the people in it." Perhaps it was the oncoming marriage that suddenly frightened him. No one knows for sure what happened that made him suddenly turn his back on the society that had entertained and amused him. There is a story that he was almost killed in his carriage when the horses bolted while crossing a bridge. The close call supposedly made him pull in his moral reins and think about saving his soul. But it is simply a story— untrue.

Gilberte wrote that his conversion was not as sudden as most people think, that he had been withdrawing from society for a year or more. At any rate, on November 23, 1654, Pascal had what is generally called a mystical experience that completely changed his life. The experience lasted for about two hours, from ten-thirty until half-an-hour after midnight. He recorded it on a piece of parchment which he then sewed into his coat and wore next to his heart. The parchment was found after his death by a servant sorting his clothes and reads in part, "Jesus Christ. I have been separated from Him; I have fled Him, renounced Him, crucified Him. Let me never be separated from Him . . . Renunciation, total and sweet."

Pascal had at last made his final choice between the world of the spirit and of science. Mathematics and physics he now regarded as "games and diversions of his youth." He dropped all his old friends, terming them "horrible attachments." When pressed to continue his work in mathematics, he said, "I can do no more. Such secrets have been revealed to me that all I have written now appears to be of little value." He wrote to Fermat, with whom he had worked so fruitfully, that "I would not take two steps for geometry's sake."

Whereas Descartes had been able to effect an uneasy synthesis between religion and science, Pascal was not. To him they were separate, and next to the problems of religion, the problems of science sank to nothingness. The saving of one's soul was a full-time job; not a minute could be spared for anything else.

After his conversion—usually called "the emotional conversion"—Pascal became closely associated with the Port-Royal convent just outside Paris. Port-Royal had a monastery attached to it, but the convent was more important. The Abbess, *Mère* Angelique, had taken her final vows at the age of nine and was made Abbess at eleven (her father passed her off as being seventeen, perhaps to relieve the burden at home where he had nineteen other children).

Pascal turned his mathematical knowledge to use by writing a textbook for the Little Schools of Port-Royal. He also acted as a missionary in his old social set and managed to convert his good friend, the Duc de Roannez, drawing him into the religious world, just as the Duc had once drawn him into the world of pleasure. De Roannez, like his spiritual advisor, completely renounced society and broke off his engagement to one of the richest heiresses in France. His family, however, was not eager to let a fortune slip through their hands simply because of what, in their eyes, was the crackpot influence of Blaise Pascal. They chose the most effective and direct solution: to dispatch Pascal immediately to the kingdom toward which he was so assiduously striving. Early one morning, their housemaid crept into Pascal's room to stab him in his sleep. But the lucky man had risen extra early and had already left for church. The de Roannezes, foiled in this attempt, never got another opportunity, for Pascal shortly moved from the hotel where he had been staying as the Duc's guest.

Once the Duc had been converted, Pascal began working on his sister, Charlotte de Roannez. With unrelenting passion and unsurpassed persuasion, he drew her, too, into the orbit of the Jansenists. "The Church has been weeping for you for sixteen hundred years," he cried. "It is time to weep for her and for us altogether and to give her all the remaining days of our lives." Charlotte capitulated before these passionate arguments, defied her mother, and ran away from home in order to become a nun. She sought sanctuary at Port-Royal-de-Paris, where she took the vow of chastity and cut her hair before her mother found her and brought her home.

There is a theory that Pascal had been in love with Charlotte, and that, being spurned, he was especially eager to convert her to Jansenism. There is no evidence, however, to support this theory. On the other hand, it rings true psychologically, for as long as Pascal remained alive, Charlotte was devout. After his death, her fervor slackened and she married a wealthy noble, living unhappily ever after.

Subsequent to his "emotional conversion," Pascal sold all his worldly goods except his Bible and other religious books. He spent most of his time living a monastic kind of life at Port-Royal. One can easily imagine him there—"a man of the world among ascetics, and an ascetic among men of the world," as T. S. Eliot described him. All his money he gave to the poor and often had to beg or borrow to keep himself alive. Martyrlike, he wore an iron belt with points on the inside so that he was in constant discomfort. Whenever he felt happy, he would push the belt's points into his flesh to bring him to his senses.

His time and talents were completely devoted to the Jansenist cause. As previously mentioned, he wrote a mathematical textbook for use in their schools, and also overhauled the teaching methods. He introduced the use of phonetics in teaching reading, and if France's little Jacques couldn't read, it was not Blaise's fault.

The severe dogmatism of the Jansenists made them the natural enemies of the more liberal orders, and a great controversy sprang up between them and the Jesuits, who tried to adapt Catholicism to the changing times in order to keep the power of the Church from crumbling even further. Pascal fought for his side by writing the *Lettres provinciales,* an apology for Jansenism that sparkles with wit and logic and that helped establish his literary reputation. Indeed, he is considered today to be as great a writer as a mathematician. The *Lettres provinciales* were written under an assumed name—though it was easy to see through the pseudonym —and printed secretely, for the Jansenists were going through a period of religious persecution, which only increased their zeal.

Pascal used the same rigorous logic in defending Jansenism as he had in constructing geometric proofs. Furthermore, he attacked mathematical and scientific truth by pointing out that the basic premises of mathematics are known intuitively, not logically and that mathematics is built on a faith in logic, just as religion is built on a faith in God. Why should one faith have more validity than the other?

Ingeniously he managed to make powerful religious weapons out of probability and gambling. "Pascal's wager" is an effective Damocles' sword hanging over the heads of doubters. What have we to lose by believing in God, he reasoned. If He does not exist, we have merely squandered a few years—a fraction of eternity— in useless good works and strivings for Grace. But if He does exist, we have an eternity of Heaven or Hell before us. Certainly, it is better to believe and be in misery for only a lifetime than to take the chance of being damned for all eternity.

If Pascal's knowledge of mathematics was superimposed on his religion, the opposite was also true. His religion made him blind to one of the greatest mathematical achievements of his day:

analytic geometry. His reason for ignoring it—or the only one that makes any sense—is that Descartes invented it. There was no love lost between the two men. Descartes had been brought up by and respected the Jesuits. To Pascal, the Jesuits were his worst enemies. Descartes stood for the supremacy of reason. To Pascal, reason was a mirage that too often obscured the face of God. And then there was the old quarrel about who had thought up the experiment to prove that air has weight.

Pascal's greatest literary accomplishment was his *Pensées*, or *Thoughts*, which he began shortly after the *Lettres*. He wrote the *Pensées* in bits and pieces, scribbling down his ideas on anything at hand, at times even using a pin to write on his fingernails when pen and paper were not handy. He died before he had a chance to finish and organize the book—it is simply a collection of ideas and feelings about men, God, life, etc. Despite its disorganized state, the *Pensées* is written in beautifully moving and graceful prose that depicts all the despair and exultation man can know: "The last act is bloody, however brave be all the rest of the play; at the end they throw a little earth upon your head, and it is all over forever." "Between us and Heaven or Hell, there is only life —the frailest thing in the world." "Time heals griefs and quarrels, for we change and are the same people no more . . . It is like a nation that we have known and meet again after two generations. They are still Frenchmen but not the same."

Obviously, Pascal was a master of prose and of expressing the most subtle nuances of human feeling—yet he denied all emotion in himself and deliberately cut himself off from every kind of intimate relationship, believing that man's innate loneliness is meant to drive him toward God and not toward other men. He would not permit himself to have friends and even his relations with his sisters turned cold and impersonal. More than once he condemned Gilberte for kissing and hugging her children. When her elder daughter was to be married, Blaise wrote his sister that she would be "guilty of one of the greatest of crimes" if she allowed the girl to marry. Marriage to Pascal was "the most base and perilous of all the conditions of Christendom," and husbands were "complete pagans in the sight of God." As a result, the wedding was canceled and his niece spent the rest of her life as a recluse, praying alone in her room and dedicating herself to God. Her sister followed the same course. Pascal had managed to save them all.

On the more positive side, he had an ingenious idea to raise money for Port-Royal. Paris was filled with narrow, twisting

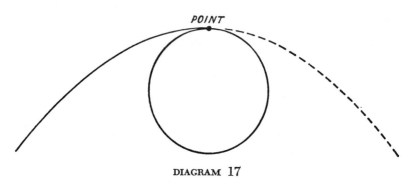

DIAGRAM 17

streets, some unpaved, the rest paved with cobblestones that oozed mud between the cracks. A pedestrian had no way of getting about except on foot or by carriage—and the latter was too expensive for any but the very wealthy. Pascal decided to start a commercial bus line. All profits would go to Port-Royal. The first route opened in 1662, and the idea of city buses spread quickly. Naturally, in his day the buses were horse-drawn carriages, but the idea is the same as for our modern motorbus lines.

Only once after his second or "emotional conversion" did Pascal give way to his interest in science. One day while lying in bed suffering from a terrible toothache, he began to think about a curve known as a cycloid when he suddenly realized that the toothache was gone. Taking it as a sign from God that He did not object to such thoughts, Blaise pursued the subject. A cycloid is the arc made by a point on a circle as the circle rolls along an

even surface (Diagram 17). Because of its many unusual properties, it is known as the Helen of Geometry, a most beautiful
curve that has given rise to several classic problems. A bridge
made in the shape of a cycloid is stronger than any other kind;
and if beads be placed anywhere on the sides of a cycloid-shaped
bowl, each bead will reach the bottom at the same time. This second example, worded in the form of a problem, was set before
the mathematicians of Newton's day.

Pascal, too, decided to challenge Europe's mathematicians with
several problems concerning the cycloid. His friend, the Duc de
Roannez, posted a large sum of money as a prize, and the challenge was issued anonymously. England's mathematicians asked
for a short extension since they had received the problem later
than France's, but Pascal's Christian generosity did not extend to
professional matters and the extension was refused. When all the
answers were in and judged, only one was found correct—Pascal's.
Almost immediately charges of theft, cheating, and underhanded-
ness began flying. Wallis, an English mathematician, accused
Pascal of having waited until all the answers were in and then
using these solutions to solve the several parts of the problem.
Wallis may have gotten this idea of plagiary from his own past
practices, for "he is so extremely greedy [of fame], that he steales
flowers from others to adorne his owne cap,—e.g. he lies at watch
. . . [and] putts downe [others'] notions in his note booke, and
then prints it without owneing the authors." Pascal publicly de-
fended himself and went so far as to accuse even his old teacher,
now dead and unable to answer, of having stolen cycloid solutions
from others. Arrogantly, he appropriated the 600 franc prize. Undoubtedly his intellectual genius enabled him to reconcile this
act with his Christian beliefs.

Three years later—in 1661—his sister Jacqueline died, and again
Pascal displayed his saintliness by reproaching Gilberte and Jacqueline's fellow nuns for grieving. The next year he himself became violently ill and refused to let anyone help him or comfort

him. "Sickness is the natural state of Christians," he said and
waved away the doctors. But a man at death's door has little to
say about his own treatment, and in accordance with standard
medical practice, Pascal was constantly bled and purged, a practice sure to cut short physical suffering by hastening one along to
the grave. Realizing that the end was near, Blaise asked to be
taken to the hospital for incurable diseases so that he could die
in the presence of the poor. (Unlike today, only the poorest people in the seventeenth century went to hospitals. People who
could afford it were treated at home.) But Pascal was too ill to be
moved. During the night of August 18, he went into convulsions
and extreme unction was administered. He regained conscious-
ness for a short while, whispered, "May God never abandon me,"
and died twenty-four hours later at one o'clock in the morning on
August 19, 1662, at the age of thirty-nine.

An autopsy was performed, but neither the cause of his years
of ill-health and suffering nor of his death was learned. Some
think he had tuberculosis of the abdominal cavity; everything
from lead poisoning and dyspepsia to softening of the brain has
been conjectured. At any rate, the life of a genius, a fanatic, a
saint, a scoundrel, and a scholar came to an early end. As a mathematician he is rated among "the first of the second rank." Everything he did of importance was either in conjunction with or
duplicated by others: Desargues and Fermat. Yet, if—and the road
to Hell must be paved with yets and ifs and buts—yet, if he had
but devoted more than the few years that he did to mathematics
and less to religion, he might stand today among the truly great.
He was well on the way to inventing the infinitesimal calculus
and probably would have if he had not had "his eyes obscured
by some evil sight," as Leibniz later said. Let us at least hope that
he won his wager.

Isaac Newton
1642-1727

The year 1642 was a momentous one. Five English colonies were flourishing in the New World; civil war broke out in Charles I's England; a blind ex-convict by the name of Galileo died in Italy at the age of seventy-eight; and the founder of modern mathematics was born on a farm in Woolsthorpe, Lincolnshire, England.

All but the last event were noted by the ever-growing, prosperous middle classes of Europe. Harriet Ayscough Newton, the recently widowed wife of a ne'er-do-well, herself belonged to this middle class. Her first child, delivered prematurely on Christmas Day, was so small and sickly that it was said he could be put into a quart mug and his "poor little weak head" had to be supported by a special leather collar. At first it was thought that the baby would die before the day was out, but the frail infant survived and was christened Isaac, after his dead father.

When Isaac was three years old, his mother remarried and moved to town to live with her new husband, the Reverend Mr. Barnabas Smith, and Isaac was left on the Woolsthorpe farm with his grandmother. At the age of twelve he was sent to school in nearby Grantham, where he boarded with an apothecary and his wife, for even a few miles was too much for the sickly Isaac to walk every day.

In school, Newton's work was on the poor side until one day he was kicked in the stomach by the class bully. Unable to avenge himself physically, Isaac decided to get even by beating the bully in his studies, and it was not long before he stood at the top of his class.

Mr. Clark, the apothecary with whom he boarded, had a step-daughter about Newton's age of whom Isaac was very fond. She was his best friend and only playmate, and their friendship lasted even into old age. "A sober, silent, thinking lad," Newton spent most of his time writing poetry, drawing pictures, and construct-ing miniature machines—clocks and windmills. He liked to fly kites to which he attached lanterns and set them aloft at night, hoping to frighten people with his artificial "comets."

At fourteen or fifteen, Isaac was taken out of school and brought back to the old stone house at Woolsthorpe to learn farming. His mother's second husband had died, she had moved back to the farm and now needed Isaac to help with the chores at home. But Isaac was more absorbed in his toy water-wheels than in keep-ing the cows and sheep from trampling through the corn. When sent to the market in Grantham, he brought along a servant to do the errands while he himself strolled over to his former lodging at the apothecary's to browse through books.

One day Isaac's uncle found him curled up under a bush com-pletely absorbed in a book on mathematics and decided that the boy was wasting his time on the farm. He persuaded Isaac's mother to send him back to school to prepare for college, so once again Isaac was packed off to Grantham and eventually, at the age of eighteen, to Cambridge University.

At Cambridge he had to work for part of his tuition, as the Newtons were relatively poor. Although he was a bright boy, there was as yet no indication of the genius that was to win him the title of one of the three greatest mathematicians of all time.

At Cambridge he put the cart before the horse and studied algebra and Cartesian geometry (analytic geometry) before learn-

ing the more basic Euclidean geometry, which struck him as being too "self-evident" and "trifling" to bother with. In 1665 he got
his Bachelor of Arts degree and began working toward his Master's.

Then, fortunately for Newton—and unfortunately for some sev-
eral thousand people who died of it—the bubonic plague caused
Cambridge to suspend classes until the epidemic had spent itself.
Newton went back to Woolsthorpe—to think and to experiment.
The next eighteen months at home were the most fruitful he ever
spent.

Using a prism he had bought at a fair, he investigated the
nature of light and discovered that it is made up of seven colors,
which when separated bend or refract at different angles. He invented the binomial theorem and calculus while studying infinite series. And, according to legend, a falling apple that landed
on his head set him to thinking about a theory of universal gravity. Although there was an apple tree which remained standing
for another 150 years next to the stone house at Woolsthorpe, it
seems rather certain that this story of the falling apple is something Newton dreamed up to dismiss the "little smatterers in
mathematics" who "bated" him. These four great achievements,
the study of the nature of light, the invention of the binomial
theorem, of calculus and of the theory of universal gravitation
—any single one of which would have won him immortal fame
—were made while he was still a young man of twenty-three!
"All this," he wrote, "was in the two plague years of 1665 and
1666, for in those days I was in the prime of my age for invention, and minded mathematics and philosophy more than at any
other time since."

Strangely enough, Newton did not tell anyone of these achievements. He went back to Cambridge, picked up where he had left
off, and kept everything to himself. His silence, however, did not
keep him from complaining in later years that no one appreciated
him or from accusing others of stealing his ideas.

In 1668 he earned his Master's degree and a year later his first recognition, limited though it was. His mathematics teacher, who considered Newton to be an "unparalleled genius," resigned so that the young instructor could take his place as Lucasian Professor of Mathematics.

Newton now made his discoveries concerning light public by using the results of his previous experiments as material in his lectures. But his highly original theories went by unnoticed by students who, for the most part, did not bother to attend the classes. It was not until 1671 that Newton's ideas on the nature of light were noticed. A reflecting telescope—the first of its kind— which he had made himself brought him to the attention of the Royal Society of London, one of the first and most famous scientific societies in Europe. A few months later Newton was made a member. Here, he thought, would be a perfect audience for his ideas. "Believe me, sir," he wrote to a fellow member, "I do not only esteem it a duty to concur with you in the promotion of real knowledge; but a great privlege, that, instead of exposing discourses to a prejudiced and common multitude, (by which means many truths have been baffled and lost), I may with freedom apply myself to so judicious and impartial an assembly." Why Newton should have had such misgivings about publishing his work earlier is a mystery, but if he expected the Royal Society to be "judicious and impartial," he was mistaken. No sooner had he submitted his paper on light than he was attacked by scholars from all over Europe, including members of the Royal Society. Newton became so annoyed that he asked the Society's secretary not to forward any more of the objectionable letters, and to his friend and future enemy, Leibniz, he wrote, "I was so persecuted with discussions . . . that I blamed my own imprudence for parting with so substantial a blessing as my quiet to run after a shadow." Gone was the dutiful feeling "to promote real knowledge." Newton resolved to keep his knowledge to himself from now on. Fifteen years went by before he could be "spurred, cajoled and importuned" to publish any major article.

During those years he continued to lecture at Cambridge, to develop his theories on gravity, and to experiment with alchemy. Like most scientists of his day, Newton believed that one metal could be changed into another. His advice to a friend about to tour the continent omitted all reference to the usual tourist attractions but mentioned "if you meet with any transmutations out of their own species into another (as out of iron into copper, out of metal into quicksilver) . . . those above all, will be worth your noting." At Cambridge he himself was literally burning the midnight oil in the course of fruitless experiment, the exact nature of which is unknown. His secretary wrote, "He very rarely went to bed till two or three of the clock, sometimes not till five or six, lying about four or five hours, especially at spring and fall . . . at which times he used to employ about six weeks in his elaboratory, the fire scarce going out either night or day, he sitting up one night and I another, till he had finished his chemical experiments." No one knows what these experiments were designed to prove. Even his secretary was ignorant of their purpose but speculated that they were "something beyond the reach of human art and industry." It is a fairly certain supposition, however, that Newton was trying to transmute metals and devoted as much time to these experiments as he did to physics and mathematics. He was, indeed, a Janus of his age, with one face turned to the future and the other, the face of alchemy and mysticism, turned toward the past.

Newton had to pay for all his laboratory materials himself—and his salary was anything but grand. The cost of supplies, coupled with his generosity, for "few went empty-handed from him," made his financial difficulties so great that he decided to resign from the Royal Society as he could not afford the shilling a week dues.

The officers of the Society, apparently realizing his straits, persuaded him to remain a member and excused him from paying dues. Another reason sometimes given for his wanting to resign is that he simply could not be bothered with attending meetings.

If Newton had resigned, his most far-reaching theory might never have been completed. Back in 1665 when Cambridge had closed down because of the plague, Newton had first conceived the idea of universal gravity but because of difficulties in calculations was unable to finish the theory until twenty years later. The exact nature of the difficulties is not known for certain. The most popular explanation is that because of a slight error made in calculating the diameter of the Earth, Newton could not make the theory agree with the facts. It was not until many years later when the diameter of the Earth was discussed at a session of the Royal Society, that Newton realized his mistake and rushed home to rework his theory. When he saw that the theory held, he became so excited that he could not finish his calculations. We can assume that he eventually managed to control himself, for he did succeed in showing that the same force that causes an apple to fall to the ground on our Earth also causes the Moon and planets to orbit around the Earth and Sun respectively. Newton called this strange force that keeps all heavenly bodies in their "appointed rounds" *gravity*, although neither he nor anyone else since has been able to explain what gravity actually is.

Yet with motion and this hypothetical gravity he could explain the universe: the Sun and stars and planets, the irregularities of the Moon's course, the tides, the equinoxes, the comets, the falling apple. The movements of all these things could be charted and described. The whole machinery of the universe could be reduced to a fictional force and three simple laws of motion:

If undisturbed, a moving body will continue to move in a straight line at a uniform speed.

A change in motion is proportional to the force causing the change and takes place in the direction in which the force is acting.

To every action there is always an equal and opposite reaction.

The first law simply means that if a ball is thrown, it will continue to travel straight ahead forever unless something stops it. The disturbing forces on Earth that cause the ball to slow down and fall are friction with the air and the pull of gravity.

The second law means that if the moving ball is struck by a baseball bat, the ball will change its motion and will move in the direction in which it is batted. If the ball is hit hard, it will travel faster than if it is bunted.

The third law means that the bat and ball act on each other. The bat stops the ball and the ball stops the bat. Since the bat is moving faster than the ball, it imparts its speed to the ball and sends it flying into the grandstands.

These laws of motion, coupled with the theory of gravitation, which asserts that all bodies are attracted to other bodies with a force directly proportional to the product of their masses and inversely proportional to the square of their distance, explain the action of every moving body. The formula for the force of gravity given by Newton is $F = kMm/r^2$, where F is the force, M and m are the masses of the bodies attracting each other, k is a constant, and r is the distance from the center of the Earth (or one body) to the center of the other body.

According to the first law of motion, the Moon should move off in a straight line. Something—some force, which Newton called gravity—was pulling the Moon out of this path. He found by observation that the distance, AB, the Moon is pulled in a certain period of time is the same as can be calculated with the formula, $F = kMm/r^2$ (Diagram 18).

The same formula can also describe the motion of objects on or near the Earth.

Newton's proof of this wonderful harmony—a mathematical harmony—in the universe revolutionized men's thinking. Formerly it had been widely believed that the Sun and Moon, the

planets and stars were made of divine matter. They were the
Heavens; Earth was only common clay. By showing that the same
laws and force operating in every man's backyard reach out and
govern the most distant star, Newton fostered the suspicion that
the heavenly bodies, like Earth, are less than divine.

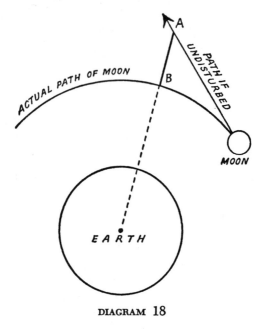

DIAGRAM 18

The one simple truth of Newton's—universal gravitation—not
only altered men's minds about divine bodies, but changed men's
minds about human minds. Through reason—the reason of science
and mathematics—Newton had been able to "demonstrate the
frame of the System of the World." He had presented people
with the most convincing proof they could want that the Cartesian method really worked.

Mathematical harmony and simplicity in reasoning became the
touchstones of all knowledge. Complicated explanations and explanations that could not be translated into numbers became sus-

pect. All truth began to be cast in mathematical language: physics, astronomy, medicine, and soon even the social sciences and arts were ruled by the Queen of Science, mathematics. By the end of the eighteenth century almost every educated man was Cartesian to the core and science and truth were one and the same to them.

That a legendary falling apple could upset the whole applecart of men's beliefs and thoughts seems amazing. Yet it happened and the world was never the same again. As for Newton, he modestly accounted for his astounding achievements by saying, "If I have seen a little farther than others, it is because I have stood on the shoulders of giants."

Among these giants of science were Copernicus, who set the Earth spinning around the Sun instead of vice versa and set the heads of his contemporaries spinning with wonder. His theory that the Earth and other planets revolve gave the Church quite a turn, for the theory upset the whole geography of Heaven. Where was Heaven now? Does it, too, revolve?

Another giant was Tycho Brahe, a Norwegian mathematician who took his geometry so seriously that he lost the end of his nose in a sword duel over an argument about a problem.

Another was his pupil, Johannes Kepler, who formulated the laws of planetary motion—that the planets move in elliptical orbits around the sun; that in their orbits the planets sometimes move faster than at other times, but that the areas of the sectors described in a given time are always equal (Diagram 19); and that there is a constant ratio for all planets between the time of revolution and the planet's average distance from the Sun; that is, the closer a planet is to the Sun, the faster it goes.

The fourth giant is Galileo who formulated the laws of motion for objects on Earth and found that every object, regardless of its weight, falls $16t^2$ feet per second, where t is any number of seconds.

The wonderful harmony in the universe discovered by these giants was made even more wonderful and harmonious by New-

ton's calculations. Kepler's laws, which were based on observations, could all be deduced through Newton's theory of gravitation. Yet this monumental theory was almost lost to the world
through Newton's reluctance to publish his book, the *Principia*,
which not only explained gravity but demonstrated its application on Earth and for all the heavenly bodies. That the book
was even written was only due to the constant prodding of

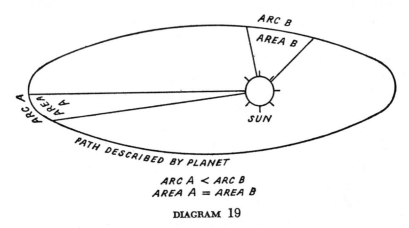

ARC A < ARC B
AREA A = AREA B

DIAGRAM 19

Newton's friend, Edmund Halley, discoverer of the famous comet
named after him. Halley accidentally discovered that Newton had
proved the inverse square law—one of the essential proofs of
gravity—and urged the mathematician to write out his findings.
Reluctantly, Newton took time from his teaching and research
to do so. At the time he was in his early forties, yet had had
his theory completed for at least five years.

No sooner had Halley seen to all the publishing details, as
instructed by a resolution of the Royal Society ("That Mr. Halley
undertake the business of looking after it and printing it at his
own charge, which he engaged to do"), than the complaints came
rolling in. Newton had purposely made the *Principia* "a hard
book" to avoid these complaints from people who knew nothing
or little about physics. Yet there were people who did under-

stand his work, and most vociferous of these was Robert Hooke who accused Newton of stealing the idea of the inverse square law. Hooke, a member of the Royal Society, was a prolific man when it came to scientific inventions and theories. Originally slated for a painting career, he had mastered Euclid's first six books in a week, discarded his artistic aspirations and turned to science. At school "he was very mechanicall, and (amongst other things) he invented thirty severall ways of flying." * The *Principia* was published in 1687, but thirteen years earlier Hooke himself had postulated the concept of gravity, writing, "all celestial bodies have an attraction or gravitiating power toward their own centers whereby they attract not only their own parts and keep them from flying off . . . but that they also do attract all the other celestial bodies that are within the sphere of their activity." Hooke, however, was unable to prove the theory, whereas Newton could. Furthermore, Newton had had the idea of gravity as far back as 1665, although he had never mentioned it. At any rate, the quarrel about who stole what from whom grew to such proportions that a bystander to the argument noted that "these two, who till then were the most inseparable cronies, have since scarcely seen one another, and are utterly fallen out."

Newton was so furious that he decided not to publish the last two books of the *Principia,* but Halley intervened and persuaded him to go ahead. The publication of Newton's treatise on his optical work, *Opticks,* however, was postponed until after Hooke's death.

The *Principia,* or *Philosophiae Naturalis Principia Mathematica,* which is the full title, brought immediate fame to its author. The

---

* This preoccupation with human flight was, so to speak, very much in the air during the late seventeenth century. One writer reported that he had been told that "a Jesuite found out a way of flying, and that he made a youth perform it . . . He [the youth] flew over a river in Lancashire, but when he was up in the ayre, the people gave a shoute, wherat the boy being frighted, he fell downe on the other side of the river, and broke his legges."

first edition sold out so quickly that a black market in the book sprang up, with copies selling for many times the bookstore price. There is a story that a Scotsman, with typical Scottish thrift, copied the whole book out by hand because he could not buy a volume at a decent price. Years later, a French mathematician asserted that the *Principia* was assured "a pre-eminence above all other productions of human genius." Newton was looked up to as the leading scientist of his day, enjoying much the same position as Albert Einstein did two hundred and fifty years later. Schoolchildren considered him to be "the smartest man in the world," even though they could not understand his theories.

After the publication of the *Principia*, Newton continued teaching and experimenting at Cambridge. Then, about five years later, when he was about fifty years old, he suffered from what has been called a nervous breakdown, often attributed to a fire in his study that burned all his notes and papers covering twenty years of experimentation. There is a record that during the winter of 1691-92, a fire did break out in his room and "when Mr. Newton came from chapel and had seen what was done, every one thought he would have run, he was so troubled thereat, that he was not himself for a month after."

Many scholars have dismissed the fire as being the cause of Newton's troubles, for his breakdown did not occur until a year and a half later when, in September, 1693, he wrote to his friend, Samuel Pepys: "I am extremely troubled at the embroilment I am in, and have neither ate nor slept well this twelvemonth, nor have my former consistency of mind. I never designed to get anything by your interest, nor by King James's favour, but am now sensible that I must withdraw from your acquaintance, and see neither you nor the rest of my friends any more."

Alarmed, Pepys wrote a common friend, asking if the great scientist were suffering from "a discomposure in the head or mind, or both." The friend answered that Newton "told me that he had writt to you a very odd letter . . . in a distemper that much seized his head and that kept him awake for about five nights . . . I fear he is of some small degree of melancholy, yet I think there is no reason to suspect it hath at all touched his understanding."

Yet a few days later, Newton wrote another strange letter, this time to John Locke, saying in part, "Being of opinion that you endeavoured to embroil me with women and by other means, I was so much affected . . . that when one told me you were sickly and would not live, I answered, ''twere better you were dead.' I desire you to forgive me this uncharitableness, for I am now satisfied that what you have done is just, and I beg your pardon for having bad thoughts of you for it . . . I beg your pardon also, for saying or thinking there was a desire to sell me an office or to embroil me." Again Newton followed up this strange letter with the excuse that "a distemper which this summer has been epidemical put me farther out of order so that when I wrote you I had not slept an hour a night for a fortnight together and for five days together not a wink."

The note of persecution in these letters cannot be explained away by Newton's notorious crotchetiness or insomnia, although even in the best of mental health he was regarded as "of the most fearful, cautious, and suspicious temper." His friend Flamsteed, with whom he later quarreled, said he was "insidious, ambitious and excessively covetous of praise, and impatient of contradiction." John Locke remarked that Newton was "a little too apt to raise in himself suspicions when there is no ground." But the events of 1693 were more than eccentricity or a "suspicious temper." Modern doctors have stated that Newton probably suffered from "a psychosis due to infectious disease with possible vitamin deficiency super-imposed on nervous fatigue due to worry."

News of his illness spread to the continent where rumors were rife that Newton had completely lost his mind. Even after he recovered, the stories persisted that he had not entirely regained his mental prowess. These stories were somewhat substantiated by the fact that Newton never again pursued an original scientific experiment. Instead, the white-haired genius—his hair had turned prematurely gray at thirty—neglected science to devote himself to governmental affairs.

Actually his "great distaste for science" dated back a couple decades. Even before he had begun the *Principia* he had written to Hooke, "My affection to philosophy [science] being worn out so that I am almost as little concerned after it as one tradesman uses to be about another man's trade or a countryman about learning, I must acknowledge myself averse from spending the time in writing about it which I think I can spend otherwise more to my own content and to the good of others."

Despite his waning interest in science, Newton could not afford to give up his teaching position at Cambridge. There was, however, a possibility that he could get a government appointment. In 1688 he had been elected to Parliament to represent Cambridge University. He apparently did not make any important contributions to the House—his only known speech consisted of asking to have a window opened. Nevertheless, he let it be known about London that he was not averse to taking a government job—which is what he referred to in his letter to Locke when he said "there was a desire to sell me an office."

Finally, in 1696, he was appointed Warden of the Mint by his friend, Charles Montague, a position which paid him a much-needed five hundred pounds a year. Four years later he was made Master of the Mint at a much higher salary. He took his government duties seriously, resigned from his post at Cambridge, and from the time of his appointment as Master of the Mint until his death, had little more to do with science.

In 1704 his second-greatest work was published, *Opticks*, the book he had waited seventeen years to release, seventeen years waiting for Hooke to die. Included in *Opticks* was the first extensive discussion of Newton's most important gift to mathematics— the calculus. Newton had known of this wonderful tool for almost forty years before he made it public, and then did so only because "some years ago I lent out a manuscript containing such theorems; and having since met with some things copied out of it, I have on this occasion made it public."

The book was reviewed by another mathematician who had also discovered the method of calculus, and in his review he implied that Newton had simply translated his, Leibniz's, works. The faithful admirers of Newton leaped to his defense and said no, it was the other way around, Leibniz had stolen from Newton. The quarrel raged on and was finally settled on the English side when the Royal Society appointed a committee to arbitrate the question. Six weeks later, the committee surprised no one by stating that "We reckon Mr. Newton the first inventor." Since Newton himself was the President of the Royal Society, Leibniz dismissed the committee's report as being filled with "malicious falsehoods." Charges of partiality issued from Leibniz's camp where the grumbling continued until 1716, when the quarrel was laid to rest with the body of Leibniz. Newton, it seems, won simply by outliving his adversary. Actually, the two men discovered calculus independently—Newton inventing it several years before Leibniz, but Leibniz publishing his work first. And according to custom, the man who first publishes or makes known a discovery or invention is given the credit—not the man who keeps it to himself for forty years or more.

What is this calculus, or Doctrine of Fluxions, as Newton called it, that had caused so bitter a quarrel? It is the algebra of the infinite, the mathematics of continuity—one of the most useful mathematical tools available.

Before discussing what calculus does and how it does it, it might be well to see how it came about, what problems inspired its creation. Calculus arose through the study of motion. Interest in motion was especially marked in the scientists of Newton's day because of their interest in the study of moving planets. The idea of motion, however, was far from new. As far back as the ancient

Greeks, a philosopher by the name of Zeno was confounding his colleagues with puzzles or paradoxes concerning motion. One of his most famous concerns a race between Achilles and a tortoise. The tortoise is given a 1,000-yard head start, but Achilles runs ten times as fast as the tortoise. Yet Achilles seemingly can never catch up, for by the time he gets to where the tortoise was, the tortoise is 100 yards ahead of him. Achilles covers this hundred yards, and now the tortoise is 10 yards ahead. Achilles runs these 10 yards, but the tortoise is still ahead by another yard. Achilles runs that yard and the tortoise is $\frac{1}{10}$ of a yard ahead. By the time Achilles runs that $\frac{1}{10}$ of a yard, the tortoise is $\frac{1}{100}$ of a yard ahead, and so on ad infinitum. Thus, logically, Achilles can never overtake the tortoise, although common sense tells us that he *will* overtake it and win the race. It can even be calculated that he will pass the tortoise at the $1,111\frac{1}{9}$ yard line, which is the sum of the converging series $1000 + 100 + 10 + 1 + \frac{1}{10} + \frac{1}{100} + \cdots$. But according to Zeno, Achilles will never get to that point, for before he gets there he must reach a previous point, and between these two points, there are an infinity of points for him to cover; therefore he can never get to the last point. There is no last point in an infinity of points.

This problem reappears in calculus in a different guise—actually it is the old problem of infinity demanding to know where he stands. Seventeenth- and eighteenth-century mathematicians did not know, so they ignored him and the play went on without infinity. They could not scrap the whole production simply because infinity was making trouble. Not until the nineteenth century was infinity given his rightful role—but that is getting ahead of the story.

One other paradox of Zeno's concerns the very essence of motion. An arrow flying through the air, he said, must at any instant be in one place—obviously, it cannot be in two places at once. While it is in this place, it is at rest there; therefore it cannot be moving. But how can a moving arrow not be moving?

Some of the force of this second argument is lost on modern

men, accustomed as we are to "stopping" motion with high-speed cameras, yet in the device of photography we can find an analogy to Zeno's reasoning. If we consider the motion in moving pictures to be continuous and not just a series of unmoving pictures being projected in rapid succession, we have what Zeno meant by motion. If the projector is stopped at a given instant, we can stop the arrow in its flight and see exactly where it is at that given instant. But now the movie is not a movie at all but a still. We have destroyed the whole concept of movies and replaced it by a still picture. The idea of stopping motion at an instant is as paradoxical and ridiculous as the idea of drying water. Little wonder that under the influence of Zeno the Greeks stuck to their finite, static, unmoving geometric figures. Although Zeno lived to be ninty-nine (he strangled himself after accidentally breaking his finger, which he took to be a sign that he had lived long enough) even a century was not time enough for him to solve his own paradoxes.

When the subject of motion was revived in the Renaissance, scientists ignored the paradoxes and abandoned the attempt to find out *what* motion was; they simply tried to measure it. Objects can move at a *constant* speed or at a *varying* speed. Constant speeds are simple things to calculate, but varying speeds present difficulties. Planets, for instance, move at varying speeds around the Sun, as does an apple falling from a tree. What, Newton wanted to know, would be the speed of the apple or planet at a certain instant?

Descartes had shown that moving lines could be plotted on a graph. As more and more points are plotted, the line determined by them moves, getting longer and longer. It is a simple matter to let the x axis stand for time and y for distance. Then the line determined by the points (x, y) indicates motion or speed. For instance, an equation, $2x - y = 0$ describes a line which might be said to be the scorekeeper for the runners, x and y, who are called time and distance, respectively. Distance and time are both moving, with distance (y) going twice as fast as time (x). If distance is in miles and time in hours, the rate of motion or speed is 2 miles an hour. This rate remains *constant*. As x changes, y changes; but their relation to each other is always the same: y is always twice as much as x. The line representing speed as well as the relation of x and y to each other, remains straight, indicating that the speed and the relation of x and y to each other are constant.

What happens when the speed is not constant, for instance, when an apple falls? Galileo showed that the apple in t number of seconds will fall $16t^2$ feet, or $d = 16t^2$. Using this formula, we can plot the position of the apple at several different seconds and connect the plotted points with a curve (Diagram 20). It is obvious from the diagram that the speed is constantly changing. How can we find what the speed is at any one instant, for example, at exactly the third second?

We can see that between the third and fourth second the apple falls from 144 feet to 256 feet, or 112 feet ($256 - 144 = 112$). Therefore, its average speed during this second is 112 feet per second. But we already know that the apple is falling slower at the beginning of the second and faster at the end. Therefore, 112 feet per second is only an *average* speed, not the *instantaneous* speed at exactly 3 seconds. We can get a closer approximation to the instantaneous speed by calculating the average speed from 3 seconds to $3\frac{1}{2}$ seconds. Substituting 3.5 for t in the formula $d = 16t^2$, we get $d = 196$ feet. Therefore, the apple has fallen $196 - 144$ feet, or 52 feet, in the half second between 3 and 3.5 seconds. We can now state its average speed during that half second as being 52×2, or 104 feet per second. Still, this is not the instantaneous speed at precisely the third second. By taking smaller and smaller intervals, we can get closer and closer to the instantaneous speed, but just as Achilles could never catch the tortoise, we can never "catch" this instantaneous speed. No matter how small an interval we take, there are still an infinite number of smaller intervals that can be taken. Table I illustrates this never-ending "progress."

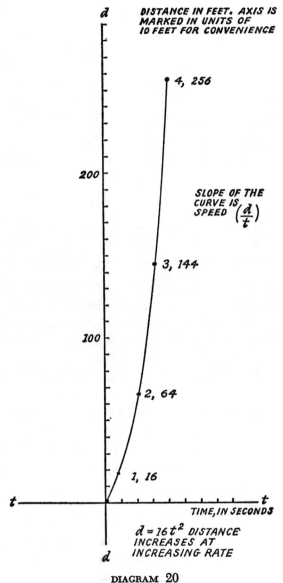

DISTANCE IN FEET. AXIS IS
MARKED IN UNITS OF
10 FEET FOR CONVENIENCE

SLOPE OF THE
CURVE IS
SPEED $\left(\dfrac{d}{t}\right)$

TIME, IN SECONDS

$d = 16\, t^2$ DISTANCE
INCREASES AT
INCREASING RATE

DIAGRAM 20

Relation of Distance and Time

TABLE I

Time			Distance			Average Speed
Beginning time elapsed (in seconds)	Final time elapsed	Difference in time elapsed (dt)	Distance covered by 3 sec.	Final distance covered	Difference in distance covered (dd)	Velocity during interval (dd/dt)
3	3.5	.5	144	196.00	52.00	104
3	3.1	.1	144	153.76	9.76	97.6
3	3.01	.01	144	144.9616	.9616	96.16
3	3.001	.001	144	144.096016	.096016	96.016

Before abandoning the idea of working with smaller and smaller intervals, let us analyze what we have done. It may make things clearer if we use symbols instead of words. Time in seconds is represented by t. Distance, which is increasing at a phenomenal rate, is *dependent* on time. Time varies and is called the independent variable. Distance also varies, depending on the variations in time. Therefore distance is the dependent variable. Distance is a function of time and this relationship can be indicated by $d = f(t)$. Thus, we have several symbols we can use to represent distance. We can call it d, or $f(t)$, or $16t^2$. For the time being let us use $f(t)$. Later we can substitute other symbols or values for this function.

The procedure as demonstrated previously for finding the average rate of speed was to find the distance traveled in a certain amount of time and to divide the distance by the time. If t represents the time at which we begin measuring the distance, another symbol is needed for the time at which we stop. The symbol we shall use is $t + \Delta t$. The symbol Δ (Greek delta) means increment; Δt means an increment in time. Now we have all the symbols necessary to record the method of finding the average instantaneous speed.

Using these symbols, let us find the average speed during the interval t to $t + \Delta t$. The distance traveled during this interval is from $f(t)$ to $f(t + \Delta t)$. To find how far has been traveled, we must subtract the initial distance from the final distance: $f(t + \Delta t) - f(t)$. The average speed for this distance is found by dividing the distance by the time elapsed. If the interval began at t and ended at $t + \Delta t$, then $t + \Delta t - t$ time has elapsed, or Δt. Dividing the distance gone by the time elapsed, we get

$$\frac{f(t + \Delta t) - f(t)}{\Delta t}.$$

This may look like a meaningless formula, but all the symbols do have meaning. For instance, $f(t + \Delta t)$ and $f(t)$ represent distance. As noted before, we can substitute other distance terms, such as $d = 16t^2$. By doing so, the formula for average speed then becomes

$$\frac{16(t + \Delta t)^2 - 16t^2}{\Delta t}.$$

(If this is not clear, you can start over using $16t^2$ for distance instead of $f(t)$. $16t^2$ will represent the initial distance and $16(t + \Delta t)^2$ the final distance.) The formula for the average speed can be simplified as follows:

Expand parenthesis: $\dfrac{16t^2 + 32t\Delta t + 16(\Delta t)^2 - 16t^2}{\Delta t}$

Collect terms: $\dfrac{32t\Delta t + 16(\Delta t)^2}{\Delta t}$

Factor: $\dfrac{\Delta t(32t + 16\Delta t)}{\Delta t}$

Reduce to lowest terms: $\dfrac{32t + 16\Delta t}{1} = 32t + 16\Delta t$

We now have a formula which gives us the average speed during the time interval Δt. How long is this time period? We can make it as short as we wish—in fact, we can make it infinitesimal. But we cannot make it zero, for if it were zero, the formula before factoring ("Collect terms") would be $\dfrac{32t(0) + 16(0)^2}{0}$,which violates the rule against division by zero. No, Δt cannot be zero, but it can be as small as we wish, and when it gets small enough, 16-times-it will also be very small. When it gets to be infinitesimal, said Newton, it will not count in the calculations. Therefore, we can ignore it. It signifies an instant and an instant is practically nothing. Thus, the average speed during an instant would be $32t + 16(0) = 32t$, and the average speed during an instant is the same as instantaneous speed. At precisely 3 seconds, the speed of the falling apple would be 32×3, or 96 feet per second, as you might have guessed from Table I. Stated as a formula, $v = 32t$. Table II shows how the velocity at 3 seconds can be found if the instantaneous speed is not known.

TABLE II

Time			Distance			Average Speed
Beginning time elapsed (in seconds)	Final time elapsed	Difference in time elapsed	Distance covered by 3 sec.	Final distance covered *	Difference in distance covered (Δd)	Velocity † ($\Delta d/\Delta t$)
3	$3 + \Delta t$	Δt	144	$16(3 + \Delta t)^2$ or $144 + 96\Delta t + 16(\Delta t)^2$	$96\Delta t + 16(\Delta t)^2$	$96 + 16\Delta t$

* Final distance covered, or distance covered after the time inverval Δt is found by substituting final time elapsed in the formula for distance: $d = 16t^2$.

† Velocity is found by dividing the difference in distance covered by the difference in time elapsed.

The similarity between Newton's method of taking ever smaller periods of time to measure instantaneous speed and Archimedes'

method of taking ever larger and smaller polygons to measure
the circumference of the circle is obvious. The only element
that Newton added—and it is the vital, necessary element—was a
symbol, Δ for *any* difference, however infinitesimally small. Everything else necessary for the invention of calculus was there in
Archimedes' time—almost two thousand years earlier.

Calculus deals with magnitudes that are changing, such as
distance. The rate at which distance changes is called *velocity*.
Calculus also deals with changing velocities, and the rate at which
velocity changes is called *acceleration*. Since velocity is a rate
of change, acceleration is the rate of a rate of change. The falling
apple, as we have seen, has a velocity of $32t$ feet a second. Its
acceleration is constant, that is, while the velocity changes from
one second to the next, it changes at a constant rate. Thus, the
acceleration is constant and equal to 32. The rate of change of
the acceleration is zero, for there is no difference in acceleration
from one second to the next. The derivatives, or differences, of
all constants are zero.

If the acceleration were not constant, we could go on to find
the rate of change of acceleration, which would be called the
third derivative and to derive a fourth from that, and a fifth from
the fourth. Velocity is the first derivative and acceleration the
second. These are the two most important derivatives in calculus.

Velocity is often called the rate of change of distance with
respect to time; and acceleration, the rate of change of velocity
with respect to time. These definitions are more precise but sometimes lead people to think that velocity means distance. Velocity
means the rate of *change* of distance with respect to time. It is the
same as speed.

The graphs for a falling apple's distance, rate of change of distance (velocity), rate of change of velocity (acceleration), and
rate of change of acceleration are shown in Diagrams 20-23 and
should help make the various distinctions clearer.

Calculus deals with constantly changing magnitudes by the

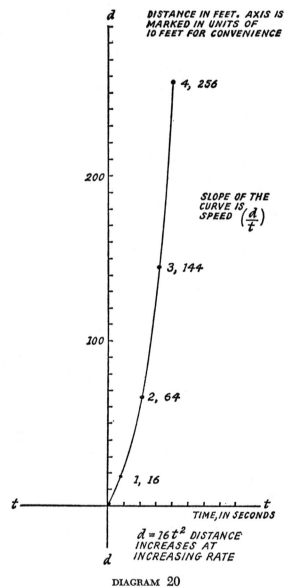

DIAGRAM 20

Relation of Distance and Time

process indicated previously of taking smaller differences between smaller and smaller intervals. This process is called *differentiating*. The reverse of differentiating is *integrating*, which is used, for example, to find distance covered when the velocity or acceleration are given.

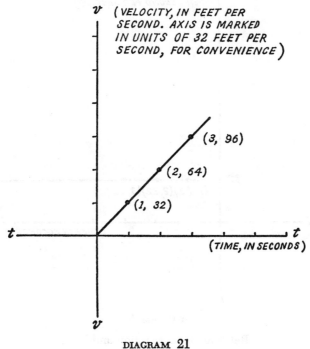

DIAGRAM 21
Relation of Velocity and Time

Calculus was developed through the study of motion, but its applications are much broader. Instead of using time and distance as the variables, heat, pressure, weight, or any number of things can be used. Nor is calculus confined to the physical sciences. It can be used to measure the effects of variables on humans or animals. These variables might be hunger, loss of sleep, drugs, etc., and would be substituted for time in the previous example.

The function of these variables, or the dependent variable, might be body efficiency, which would correspond to distance in the falling apple problem. Instead of measuring speed the rate of change of efficiency would be measured.

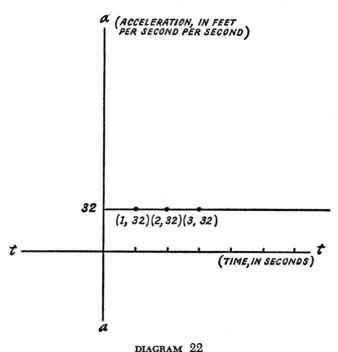

DIAGRAM 22

Relation of Acceleration and Time

Calculus from the start proved itself immensely useful; but it rested on an unsound, pragmatic foundation rather than one of pure logic. The whole idea of dismissing infinitesimal values or instants as not worth bothering about is rather cavalier for so precise and logical a science as mathematics. Mathematicians were alarmed, but what could they do?—calculus gave correct results.

Thirty years after Newton published his *Opticks* containing

the Doctrine of Fluxions, Bishop Berkeley wrote a knowledge-
able and extremely witty attack:

> Now, as our Sense is strained and puzzled with the percep-
> tion of objects extremely minute, even so the Imagination,
> which faculty derives from sense, is very strained and puzzled

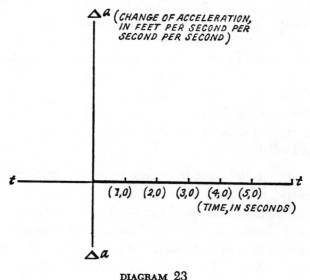

DIAGRAM 23

Relation of "Change of Acceleration" and Time

to frame clear ideas of the least particles of time, or the least
increments generated therein: and much more so to comprehend the moments, or those increments of the flowing quantities
in *statu nascenti*, in their very first origin or beginning to exist,
before they become finite particles. And it seems still more difficult to conceive the abstracted velocities of such nascent imperfect entities. But the velocities of the velocities—the second,
third, fourth, and fifth velocities, etc.—exceed, if I mistake not,
all human understanding. The further the mind analyseth and
pursueth these fugitive ideas the more it is lost and bewildered;

the objects, at first fleeting and minute, soon vanishing out of sight. Certaintly, in any sense, a second or third fluxion [derivative] seems an obscure Mystery . . . And what are these fluxions? The velocities of evanescent increments. And what are these same evanescent increments? They are neither finite quantities, nor quantities infinitely small, nor yet nothing. May we not call them the ghosts of departed quantities? . . . All these points, I say, are supposed and believed by certain rigorous exactors of evidence in religion, men who pretend to believe no further than they can see. That men who have been conversant only about clear points should with difficulty admit obscure ones might not seem altogether unaccountable. But he who can digest a second or third fluxion, a second or third difference, need not, methinks, be squeamish about any point in divinity . . .

The difficulty inherent in infinitesimals was ignored by most mathematicians. Calculus gave the right answers and a little thing like infinitesimals could not be allowed to wreck the wonderful tool. Besides, were infinitesimals any worse than irrational numbers? Irrationals were used constantly, and their use made them accepted. Yet they defied the rules of arithmetic in that they could not even be added. Mathematicians took the attitude that he who can digest an irrational need not be squeamish about any point in calculus.

The problem of infinitesimals, or the infinitely little, is the same as that of infinity, or the infinitely great. In one case the numbers approach zero; in the other, they approach infinity. In neither case do they ever reach their goal. The problem of the infinitely great or little exists in every area of mathematics. It is no surprise to find it in calculus, where it appears in other guises besides infinitesimals. For instance, in the problem of the falling apple, at each second the apple is in a certain place—but for each place it is in, is there a certain time that has elapsed? The commonsensical answer is yes. The seemingly logical answer is no. For example, after one second the apple is at the 16-foot line; after two seconds, at the 64-foot line; after three seconds, at the 144-foot line. Thus, we can go on and on accounting for every minute, but there are great gaps in the corresponding distances. There are obviously more times that the apple can be in places than there are places that the apple can be in in times! This paradox, like all those arising from infinity, was not cleared up until the middle of the nineteenth century. Yet the answer was staring mathematicians in the face long before that. They simply refused to recognize it.

The preceding describes Newton's Doctrine of Fluxions—today known as calculus—which was published in 1704. By that time Newton had been living in London several years, having left Cambridge in his early fifties. With thirty years of life still ahead of him, he had turned his back on his field of greatness to devote himself to the Mint. The first third of his life had been devoted to education and the formulation of his theories; the second third to the development, consolidation, and publication of those theories; the final third to reaping the honors.

His appointment to the Mint represents the traditional English practice of honoring brilliant men by giving them government posts—a practice that probably accounts for England's abundance of outstanding statesmen. At the same time, it skims the cream off other professions, leaving only the second-rate to solve the problems of science and industry. In Newton's case, the loss to science was undoubtedly greater than the gain to government. Yet, by his own admission, he had long been tired of science and it is, therefore, unlikely that he would have made any further major contributions.

It would be a mistake to assume, however, that Newton changed radically after Cambridge or that he lost his intellectual drive. True, science became less important to him, although he was still capable of solving the most difficult problems. Even at the age of 74, when he returned home from the office one evening, he tackled an extremely difficult problem set forth by Leibniz as a challenge to Europe's mathematicians. Newton solved it before he went to bed. His ability of intense concentration never left him. "I keep the subject constantly before me," he explained, "and wait till the first dawning opens little by little into the full light." It was this almost supernatural power of concentration— a power perhaps never equaled—that was one source of his genius.

His second honor came in 1703 when he was made President of the Royal Society, a position which he held to his death. Two years later, at the age of sixty-two, he was knighted and began to move in royal circles. Queen Anne declared that she was honored to be alive at the same time as so great a man as Newton. Sir Isaac, however, thought of himself as "a boy playing on the seashore, and diverting myself in now and then finding a smoother pebble or a prettier shell than ordinary, whilst the great ocean of truth lay all undiscovered before me."

In the last third of his life, Newton continued to collect his pretty pebbles on the beach of truth. But these pebbles were now religious and historical ones instead of scientific. His interest in theology and history was of long standing. At Cambridge he had studied history for relaxation, and his interest in religion encompassed all others. Even his mathematical work was, in a sense, religious. Newton believed that there was an order in the universe and did not doubt that God must have put it there. Like Descartes, he unconsciously used God as both premise and conclusion in his reasoning. He began with the idea that a simple, mathematical universe had to be made by "a cause that understood." The theory of gravity proved that the universe was simple and mathematical; *ergo*, God exists. Scientific investigation to Newton was an investigation of God's works, an uncovering of the wonderful design in the universe; and every piece of truth discovered was additional proof of His existence.

When Newton's friend, Halley, who apparently inclined toward atheism, made frequent derogatory comments about religion, Sir Isaac would chide him gently, saying, "I have studied these things, —you have not." Newton was thoroughly religious, devout, pious, mystical—and heretical. He was a Unitarian; the Church of England believed in the Trinity. Yet few people knew of his unorthodox beliefs. The only indication he ever gave of heresy was in his refusal to take Holy Orders, the customary practice of university professors. At other times his courage was not as great. He never uttered a word in behalf of friends who were punished for holding the same heretical views. Newton was obviously no Luther displaying his religious objections in the public square; he was a quiet, retiring man who stored his secret in his heart and kept his written ideas on religion hidden away in a large box that he brought with him from Cambridge.

In his old age he combined his knowledge of history and religion by writing *The Chronology of Ancient Kingdoms Amended,* which puts ancient events—including those described in the Bible —in chronological order. A tremendous amount of painstaking work went into the project. A contemporary reported that Newton "wrote out eighteen copies of its first and principal chapter with his own hand, but little different one from the other." All in all, Sir Isaac set down over a million words on theology. In theology and history, as in science, he absorbed himself completely.

Like anyone else, Newton was not immune to the pleasures of honor and financial security. The common supposition that genius is content to be decorated with pearls of wisdom and to dine on food for thought is certainly not true in his case. It has even been suggested that his weariness with science was more a case of his being tired of being poor. For this reason he may have sought and accepted the position at the Mint.

In London, Newton's finances improved tremendously. Whereas at Cambridge his income had been merely adequate, in London he prospered enough to buy his own horses and carriage and to

afford a staff of six servants. At his death he left a considerable
fortune of some 32,000 pounds.

As for honors, Newton found those accruing from academic
work to be bitter. Constant "feuding and warring in print" greeted
the publication of all his epochal works. No, neither academic
salaries nor academic honors could compare with those in other
areas. Having once tasted worldly honor, he was eager for more.
At the age of sixty-three, he tried to join the English aristocracy
by applying for a coat of arms, a distinction held only by people
who could trace their ancestry to a noble, however obscure.
Newton's attempt to erect an aristocratic Scottish background
for himself failed, however, and he had to be satisfied with the
honors won by purely personal accomplishments.

Despite this setback, Newton's life was pleasant and well ordered. He worked hard at the Mint and returned to a well-run
house, presided over by his widowed niece, Catherine Barton.
Her wit, charm, and good looks attracted scores of bachelors to
the Newton home, one of whom she eventually married: John
Conduitt, a man nine years her junior who later succeeded Sir
Isaac at the Mint. Catherine was indispensable to Newton, who
was the epitome of the absent-minded professor. At Cambridge
he had frequently forgotten to eat or even to dress himself.
Sometimes he went to bed with all his clothes on. When he
worked—often for eighteen- or nineteen-hour stretches—he forgot
everything but the problem at hand. There is an anecdote that
once when dinner was served to him, he completely forgot about
it and wandered off, deep in thought. A visitor came into the room
to wait for him. Growing hungrier by the minute, the visitor
spied Newton's untouched dinner and, like Goldilocks, sat down
and ate it all up. When Newton returned and saw the empty
plate, he remarked, "How absent we philosophers are. I really
thought that I had not dined."

His secretary noted that Newton "ate very sparingly, nay,
ofttimes he has forgot to eat at all . . . He very rarely went to

dine in the hall except on some public days and then if he has not been minded, would go very carelessly with shoes down of heels, and his head scarcely combed."

Despite this neglect of his person and health, which he constantly attempted to bolster with homemade medicines and a special breakfast tea consisting of boiled orange peels, he was untroubled by any major physical disorders until his very old age. His mind remained keen to the end, but his memory began to fail. He kept his hair, all his teeth, and his eyesight.

At the age of eighty, he began to be bothered by terrible pains which were diagnosed as "the stone," most likely kidney stones. Three years later he developed the classic symptom of good living, gout, and turned over most of his work at the Mint to Mr. Conduitt, his nephew-in-law. He continued to preside at the Royal Society, but more than one member noted that Sir Isaac often fell fast asleep at meetings.

When he was eighty-four, his pains from "the stone" returned with such a violence that "drops of sweat ran from his face." This diagnosis of "the stone" has been recently discredited, and it is believed that he had angina pectoris. Sir Isaac Newton died on March 20, 1727, at the age of eighty-four.

His body lay in state at Westminster Abbey, where he was buried. His grave there is marked by a plain stone inscribed *"Hic depositum est Quod Mortale fuit Isaaci Newtoni"*—Here are deposited the mortal remains of Isaac Newton. A few years later his heirs erected a monument to him which was placed in a favored spot of the Abbey, a spot that had long been denied to high-ranking nobility.

Leonhard Euler
1707-1783

The erroneous but popular conception of mathematicians as an eccentric lot is hardly refuted in the persons of Cardano, Pascal, or Newton. As far as sheer conceit, fanaticism, or crotchetiness go, these three are perfect examples. But mathematicians are, basically, human beings, with the same failings and virtues as any other group of humans. If the three above were slightly eccentric, dozens of others were Mr. Normal himself. Leonhard Euler is a case in point. In manner and appearance he gave the impression of being a small-town, middle-class minister rather than Europe's greatest mathematician—an impression that had some basis in actuality, for Euler not only came from a small town but was originally slated for a clerical career.

Born in Basel, Switzerland, on April 15, 1707, of Marguerite and Paul Euler, he grew up in a tiny outlying town where his father was the Calvinist minister.

Improbable as it may seem, this small-town boy found his way into some of Europe's most dazzling courts during those glory days of kings—Louis XIV, Catherine and Peter the Great, Frederick the Great of Prussia. All had their own Versailles or replicas thereof, and all vied with each other in cultivating the arts and sciences. In the eighteenth century there were no kings of commerce to support scientific research, education, art, and other worthy causes. This philanthropic role was played by kings of the land, who skimmed the intellectual cream off Europe for their academies and courts.

Euler received his first schooling from his father, who had studied mathematics under one of the Bernoullis, a famous and talented family that produced more than a dozen top-rate mathematicians. While still a young boy Euler entered the University of Basel where he, too, studied under a Bernoulli, Johannes, the brother of his father's teacher.

At Basel his favorite subject was geometry, but after getting his Bachelor's degree at fifteen and Master's at sixteen, he dropped mathematics to study theology and Oriental languages in accordance with his father's wishes. With a little pressure from the Bernoullis, however, Paul Euler soon gave into his son's pleas to return to the study of mathematics.

At eighteen, Euler published his first mathematical paper, a treatise on the masting of ships, which he submitted in the annual contest held by the French Academy of Science. Although he was competing against Europe's top mathematicians and scientists, many of them two or three times his age, he won second prize. His paper is unimportant as far as the history of mathematics goes, but it does illustrate one aspect of Euler's method. He never checked his results experimentally. Coming from landlocked Switzerland, he knew next to nothing about ships or their sails. But this lack of firsthand experience did not bother him, for since his conclusions on the height and thickness of masts were "deduced from the surest foundations in mechanics; their truth or correctness could not be questioned."

While Euler was still a schoolboy engrossed in his studies, the free-wheeling, swashbuckling Tsar, Peter the Great, made a tour of Europe. He avoided the fashionable resorts and watering places and concentrated instead on dockyards and universities. Not satisfied by just looking, the six-foot-seven-inch Tsar picked up a hammer and saw and joined the workmen in building ships. In Holland he took a course in anatomy and became the first—and probably only—grand monarch to pull teeth and perform surgery. Everything he saw convinced him more and more of how backward his own country was in comparison with the West, and he returned home determined to lift Russia to the level of France and England. French became the language of the Russian court; nobles were told to shave their outlandish beards; and ladies were introduced to French fashions. Peter built himself a miniature Versailles and a new capital, St. Petersburg, later called Petrograd, and then Leningrad. He founded a medical school and put his own learning of anatomy to use by performing an autopsy on his sister-in-law. And he drew up plans for an Academy in St. Petersburg, which were carried out by his widow, Catherine I.

Two of Euler's friends, Daniel and Nicholas Bernoulli, were invited to come to Russia to teach and study at the new Academy. They went and wrote back about an opening in the physiology department, urging Euler to brush up on the subject and apply for the position. At their own end, they persuaded Catherine I that they knew just the man for the vacancy—Leonhard Euler. Euler quickly crammed in a few courses in medicine and anatomy, applied for the position, was accepted, and at the age of twenty transplanted himself to St. Petersburg. Because of some sort of disorganization or other inscrutable bit of administrative red tape, he was unexpectedly given a position in the mathematics department.

The beginnings of the Academy were inauspicious, if not makeshift. Each professor had to bring with him two students—a policy that resulted in the Academy's opening with twenty teachers and forty pupils.

In addition to the scarcity of pupils, there was a like scarcity of money. Catherine I had died the day Euler reached Russia and her successor's regent, less enlightened about education, considered the Academy an unnecessary drain on the royal exchequer. Within three years Euler found himself on the verge of bankruptcy. He considered giving up his position to join the Russian Navy and was on the verge of doing so when the young Tsar died and was replaced by the Empress Anna, who was looser with the purse strings although tighter with the whip.

Euler's friend Nicholas Bernoulli died during his first years in Russia and Euler succeeded him as head of the Natural Philosophy department. By 1734 his finances had improved to the point where he could afford to marry. He took as his wife Catharina Gsell, the daughter of a Swiss painter enticed to St. Petersburg by the late Peter the Great. Continuing his work in mathematics, he flooded the scene with writings, many of which attracted a good deal of attention. At the age of twenty-five his first major work—a two-volume book on mechanics—was published and the author hailed as a genius. Several other papers were submitted to the French Academy where he and the Bernoullis seemed to have cornered the market as far as prizes went. Between them, they walked away with twenty-eight prizes, Euler winning twelve.

Quantity as well as quality characterized Euler's work. Undoubtedly, he is the most prolific mathematician who has ever lived. Writing incessantly—short papers, longer treatises, books, all of which required a tremendous amount of study and thought—he produced enough to fill over ninety large volumes. His mind worked like lightning and was capable of intense concentration. But unlike Newton, Euler needed neither quiet nor solitude. Most of his work was done at home in the bedlam created by several small children noisily playing around his desk. Euler remained undisturbed and often rocked a baby with one hand while working out the most difficult problems with the other. He could be interrupted constantly and then easily proceed from where he had left off without losing either his train of thought or his temper.

Research, teaching, writing and his growing family filled his days, but the constant reading and work took their toll. The year after his wedding he received a problem from the French Acad-

emy and, devoting three days of intense concentration to it, found
a solution. Other mathematicians asked for several months to
solve the same problem. Euler's three days' work "threw him into
a fever which endangered his life" and, according to the medical
diagnosis of the time, was the cause of his losing the sight in his
right eye.

At the age of thirty-five, Euler was invited by Frederick the
Great of Prussia to come to Berlin to teach and do research. He
accepted readily, happy to leave a Russia which was being
bathed in blood through the relentless efforts of Empress Anna
to clean out spies and traitors.

Frederick, like Peter the Great, was determined to aggrandize
his country through art and science, as well as war. A much
more polished man than his Russian counterpart, Frederick had
managed to acquire a literary veneer despite the hindrances set
up by his father. Frederick's father, an ogreish Philistine if ever
there was one, considered books and culture as effete and useless
and was constantly popping his son into a uniform and thrusting
a sword in his hand to make a man of him. Nevertheless, when
his father died, Frederick reverted to a cultural life and immediately invited Europe's leading intellectual lights to his court.
Voltaire, his idol and "the finest ornament in France," was one
of the first to be asked. Frederick tried to get the Bernoullis as
well as Euler, but failed.

At Berlin Euler's reputation rose higher than ever through the
publication of his *Introductio in Analysis infinitorum,* and *Insti-
tutiones calculi Differentialis.* "Through them," said a later
scholar, "he became the mathematics teacher of all Europe." The
calculus, which had been invented—or at least made public—only
a few decades before, was still largely undeveloped. Along with
many other mathematicians, Euler believed that it was not a
perfect tool but gave correct results because the errors offset each
other. Fortified with this belief, he struggled to clarify and simplify the analytical operations. His two books on calculus were

not only the best, but the most understandable of the time. Strange as it may seem to us today, they enjoyed a tremendous popularity despite their technical subject matter. The eighteenth century was still an age when no man could consider himself educated without a knowledge of mathematics, for on mathematics all knowledge was based. Its methods set the standard and became the model for every other branch of learning. Indeed, the belief was prevalent that everything—any idea or fact—could be summed up mathematically. Leibniz had even started a grand scheme whereby all ideas were to be reduced to symbols, which could then be handled in the same way as algebraic symbols. This scheme, he believed, would rid the world of wars, for all disputes and differences could be settled peacefully and fairly by simply juggling symbols. The belief that wars are caused only by injustice may be a bit naive, but the general plan of using symbols for ideas is not as farfetched as it might seem. Today an artificial language capable of expressing detailed ideas is being developed for use by electronic computers and a whole branch of modern mathematics, symbolic logic, has been erected on the ruins of Leibniz's scheme.

In Berlin, Euler continued to produce his papers—some being sent to Russian journals, for he still received a salary from the Academy of St. Petersburg; and some being published by the Berlin Academy. The papers fairly gushed forth. He calculated as easily "as men breathe, or as eagles sustain themselves in the wind"; his photographic mind skimmed over the most difficult problems and came up with solutions in a flash—throwing "new light on nearly all parts of pure or abstract mathematics."

Euler worked in every field of mathematics: analysis (he was called "analysis incarnate"), algebra, geometry, and number theory. In all of these branches he consolidated and united the work that had been done before. He supplied the missing links and went on to develop the unfinished theories of others. He extended the applications of analytic geometry, for instance, to three dimensions, where he found general equation forms for planes and the different solids. A line, as shown in the chapter on Descartes, is of the form $ax + by + c = 0$. In the three-dimensional system, the general equation for the corresponding figure, a plane, is $ax + by + cz + d = 0$; the general equation for a sphere, $x^2 + y^2 + z^2 = d^2$, also bears a resemblance to that of the circle. Compare it with $x^2 + y^2 = c^2$, the equation of a circle.

If the general equation in geometry (plane) has two variables (x and y), and if three-dimensional geometry (solid) has three variables (x and y and z), then four-dimensional geometry should have four variables, and n-dimensional geometry n variables. At least that was the thought that struck some mathematicians a hundred and fifty years later. True, there is no fourth dimension, realistically speaking, but there is one *mathematically* speaking, for we can give equations that describe a figure in four dimensions. Today this fourth dimension is usually taken as *time*—and who has not heard of its use in Einstein's theory of relativity? Later it will be shown how this idea of four dimensions had to be combined with something else before it could be used to describe the universe and its many time-space worlds. Einstein could never have formulated his theory without building on what others had done. Like Newton, he, too, stood on the shoulders of giants: Descartes and Euler were two of them.

Euler was one of hundreds who tackled the problem known as "Fermat's last theorem": to prove that there are no positive integers such that $a^n + b^n = c^n$, where n is more than 2. A problem of this type falls into the category of number theory, a branch of mathematics where the problems are so simple that even an amateur can understand them, but where the solutions are so difficult that they tax the best minds. For instance, the example above contains the *particular* problem of proving that no positive integers exist, the sum of whose cubes equals another cube. We know that there are many integers whose squares added together equal a square ($3^2 + 4^2 = 5^2$; $5^2 + 12^2 = 13^2$, etc.), but what about cubes? No one has ever found two cubes equal to a third, but that does not *prove* that they do not exist. Of course the problem is even more general. It involves proving that there are no two integers which raised to *any* power above the second can be added together to produce a third number of the same power.

The problem was—and still is—especially intriguing because of the following notation made by Fermat in the margin of a book: "It is impossible to partition a cube into two cubes, or a biquadrate [fourth power] into two biquadrates, or generally any power higher than a square into two powers of like degree. I have discovered a truly wonderful proof of this, which, however, this margin is too narrow to hold." Because Fermat lacked a few inches of space, mathematicians have wasted reams of paper in trying to find this "truly wonderful proof." Nobody has ever found it. Euler set his mind to it, too, and failed, although he did succeed in proving the impossibility when *n* equals 3 or 4.

Euler was the great organizer in mathematics, and in organizing the subject, he supplied many missing links. His work in analytic geometry has been mentioned. In trigonometry, he invented the calculation of sines and put the whole subject on an algebraic rather than a geometric basis. Until he came along, trigonometry consisted of a number of unrelated formulas which even as far back as Archimedes' time were used to find the lengths of the sides of triangles. Trigonometric functions had been invented but not recognized as ratios. Euler tied the whole thing together into one consistent whole—just as it is taught in schools today.

Unfortunately, Euler did not confine his writing to mathematics. Against the advice of his good friend Daniel Bernoulli, he wrote several pieces on religion—after all, he had started out to be a minister and probably felt he knew something about the subject. "His piety was rational and sincere; his devotion was fervent," as one of his students described it, but his arguments were poor and were easily cut down by "that most terrible of all the intellectual weapons ever wielded by man, the mockery of

Voltaire." Voltaire, who held the place of honor at Frederick's court, was Europe's greatest wit and most articulate ridiculer of organized Christianity. His acidulous tongue had twice landed him in the Bastille and his ideas, "scandalous, contrary to religion, to morals, and to respect for authority," had won him the undying enmity of the Church.

What were these ideas that Euler and Voltaire batted back and forth? They were ideas that had sprung from mathematics and physics and concerned Truth, God, and His importance.

From the beginning of time, man has attempted to order his world. Storms, eclipses, floods, droughts, sickness occur suddenly, inexplicably, and seemingly without cause. Yet even ignorant, uncivilized man found this haphazard behavior of nature repugnant. Therefore, he made up explanations. It rained because the sky was unhappy, or it didn't rain because the sky was happy. Gradually this animism gave way to a more sophisticated concept. Gods ran the universe and, depending on their moods, dispensed disaster or prosperity. Naïve as these beliefs may seem to us, to our ancestors they were the only way in which to explain the otherwise inexplicable chaos of the world. In time these ideas were refined even further. One God instead of many was sovereign and He ruled not by whim but by law based on His infallible knowledge. These laws of God, which mortals could never hope to understand completely, attained the status of Truth.

Science is the newest way of explaining nature's behavior, of bringing order out of chaos by finding a cause-effect pattern in phenomena. Yet, with the advent of science, men did not completely abandon their gods. Instead, they blended religion and science. Every scientific discovery, every added instance of a mathematical order in the universe, was only added proof of the existence of an intelligent Creator who had put the order there. Newton, for instance, firmly believed that in studying the laws of the universe he was studying God.

A hundred years later—by Euler's day—the emphasis on religion had waned and in many quarters the scientific explanations alone were considered sufficient. Science, formerly the servant of religion, had ousted his master and assumed for himself the title of Arbiter of Truth.

Yet God could not be dispensed with entirely; He had to be retained to explain what science still could not, namely, the origin of motion and matter. That was His only function. He had become an impersonal, indifferent, passive being, a "great watchmaker," as Voltaire called him, who had simply wound everything up and set it running. By listening to the tick-tock-ticks of the gigantic watch, measuring and timing them, scientists could determine what pattern they followed and could comprehend the mechanism of the universe. So successful were they in their observations that many came to believe that the only valid knowledge—indeed, the only possible knowledge—is that gained from observation. Since God cannot be observed, said these Empiricists, we cannot say He exists, and they looked on Voltaire as an antiscientific bigot because "he believes in God." Yet in Euler's eyes Voltaire was a heretic whose "religion" bore little resemblance to Christianity. Free will, for instance, one of the cornerstones of the Christian religion, was scoffed at by Voltaire. If all things—planets, stars, falling objects, everything in nature—act according to certain laws, why, he asked, should "a little animal, five feet high . . . act as he pleased." Man is simply another moving cog in the whole vast machinery. His freedom is only illusionary. As Schopenhauer said, *"Der Mensch kann was er will; er kann aber nicht wollen was er will."*

The theological explanations Euler offered to reconcile Voltaire's ordered, deterministic universe with man's free will fell on deaf ears. Almost everyone at Frederick's court—including Frederick himself—professed a belief in scientific deism. Indeed, deism was becoming the official religion among the educated classes everywhere. Benjamin Franklin and Thomas Jefferson, contem-

poraries of Euler, both were deists. The Declaration of Independence mentions "the Laws of Nature and of Nature's God," referring
to the mechanical laws of the universe and the impersonal God
who laid them down. The Constitution of the United States, written when deism was at its height of popularity, makes no direct
mention of God at all, nor does the oath that the President must
take before he enters office.

Despite the popularity of Voltaire's ideas, Euler refused to
budge from his position as a good Calvinist. He lost all the battles of words but continued to iterate his belief in a Christian God
and to lead his family in prayer every evening.

And yet, compelling as Voltaire's arguments are, the whole concept of determinism is riddled with flaws, many of which bear
on or later appear in mathematics, as will become evident.

If, as Voltaire maintained, man is part of an ordered, deterministic universe, then all his ideas are also determined. They are imposed on him in the same way that it is imposed on the stars to
keep to their orbits in the heavens. Man cannot reject ideas as
"false" or accept them as "true," for it is also determined whether
he will reject or accept them. Truth then, cannot be discovered
by the human mind. What appears to us as truth is merely what
the machinery of the universe has ground out for each of us. Thus,
the ironical but logical outcome of this rationalistic philosophy is,
in the end, to discredit man's reason and his ability to determine
truth.

Furthermore, if the doctrine of determinism is true, it contains a fatal paradox. Of the almost infinite number of ideas imposed on man, he can never know which are true or which are
false. If he could know then the idea would not be imposed. Yet
man does know the truth of one idea: that ideas are imposed.
Therefore, by granting that all ideas are imposed, man must come
to the logical but absurd conclusion that all ideas are not imposed!

A third objection can be made against man's ability to find a
pattern in determined, mechanical phenomena. And the objection

was made, made, and largely ignored. When Euler was still in Russia, a young man by the name of David Hume had come along and struck science a blow from which, according to Bertrand Russell, it has never recovered.

At the age of twenty-six, Hume finished his *Treatise of Human Nature*, in which he shows that it is impossible to find any patterns at all by observing nature. What right have we to assume, he asked, that there is any connection between events? It does not logically follow that because we put a pot of water on the fire the water will boil. We *expect* that the water will boil and this expectation, according to Hume, has led us to believe that there is a *necessary* connection between heat and boiling water. This "necessity is something that exists in the mind, not in objects." Just because one event is always followed by another event does not mean that the first event *caused* the second. And just because these two events have always occurred together does not mean that they always will. "The supposition, that the future resembles the past, is not founded on arguments of any kind, but is derived entirely from habit." Thus, if Hume's objections are accepted, there can be no such thing as scientific truth; science becomes nothing but a pile of unrelated data.

The arguments to meet these different objections were many and varied and for the most part weak. Religiously inclined men, such as Euler, attempted to show that man is exempt from determinism, that he is different from other animals. This difference lies in the assumption that man has a soul. Unfortunately, this consoling difference appeared less tenable when Darwin announced that man had evolved from the same ancestors as apes. Exactly when and why man, and not apes, should have acquired a soul is a question impossible to answer and futile to ask.

Another tack was to put the whole paraphernalia of order into the human mind, to admit that order and scientific laws are simply the creation of man, who cannot think chaotically. He can never know whether order exists in reality or only in the eye of the beholder. He can only know his own mental ordering and for him that is Truth. Man must order on a cause-and-effect basis because it is the way his mind works. He must assemble events in space and time because he cannot do otherwise.

The force of this argument is undeniable and remained so until the twentieth century, when physicists showed that man does not necessarily have to conceive of all events as having a cause and effect or of happening in space and time. A whole microcosmic world exists in which scientists have been not only unable to find a causal pattern but have discovered indications that such a pattern does not exist. Furthermore, in Einstein's theory of relativity, events do not occur in space and time; they occur in space-time. So now where is the so-called order that the human mind imposes?

The whole doctrine of determinism today is enjoying a revival of interest; as Werner Heisenberg wrote recently, "Many of the abstractions that are characteristic of modern theoretical physics are to be found discussed in the philosophy of past centuries. At that time these abstractions could be disregarded as mere mental exercises by those scientists whose only concern was with reality, but today we are compelled by the refinements of experimental art to consider them seriously." * The question of determinism and its corollary, causality, may never be answered. It may rank with the other vast riddles of the universe—yet it is the fundamental assumption of the most rigorous method for finding truth that the world has ever known.

Two other anti-anti-determinism arguments must be mentioned, not for their effectiveness but because of their effect. Both are essentially evasions. The first is to admit that reason is incapable of finding truth and to substitute for reason and truth something else—emotion, instinct, passion, faith. These irrational forces are what men should obey rather than ineffectual reason. Pleasure,

---

* Werner Heisenberg, *The Physical Principles of the Quantum Theory*, published by the University of Chicago Press. Copyright 1930 by the University of Chicago.

power, salvation, or any other goal are what men should strive for rather than unknowable Truth. The Romantic movement of the late eighteenth century is one example of an irrational philosophy at work; the Nazi movement of the early twentieth is another.

The final argument is one of "common sense" and pragmatism. Scientists met the issues obliquely by saying, "Well, everything including my ideas may be determined; the order I think I see may be imposed or self-manufactured. It doesn't matter. I'll just look around anyway and describe mathematically what I see. I don't claim that these descriptions are Truth or that they accord with what actually exists. They are only descriptions of my observations. As I see more and more I can adjust my mathematical descriptions to handle these added observations. Call my descriptions what you want—truth, probable truth, science, nonsense. I can only say that they are very useful and until someone finds a better way, I'll continue with mine." Thus, truth for the scientist is dynamic rather than a static absolute. It is constantly being uncovered (or created), refined, and developed so that consistency and simplicity can be maintained. The scientist makes no absolute claims for the veracity of his findings, only probable ones. He has abdicated the throne where he reigned as Arbiter of Truth and is now employed as an administrator in the same court. Science is no longer a doctrine, it is a technique.

And what about mathematics? Do its truths exist independently of man, and if so, how much is accessible to human reason? Or has man himself created mathematics whole cloth out of the fabric of his mind, weaving together postulates with logic? Is mathematics, after all, only a technique and not absolute Truth? Neither Euler nor Voltaire nor anyone else thought of asking these questions. Mathematics had been synonymous with Truth for so long that its position was unassailable. It was its offspring, science, that had to meet these philosophical hurdles—hurdles have been given in some detail because later mathematics, too, had to meet them, and used the same arguments against the same objections.

The contests between Euler and Voltaire, fought on the battlefield of religion and science, were later transferred to the battlefield of mathematics. Euler did not live to see the second conflict,
which was just as well, for he had enough arguments in his lifetime to keep him busy—arguments that he always lost. "Our
friend Euler," wrote one of his admirers, "is a great mathematician
but a bad philosopher . . . It is incredible that he can be so
shallow and childish in metaphysics."

Antagonistic religious beliefs were not the only thing that
plagued Euler in Berlin. The members of the Academy indulged
in constant intrigues and feuds—and Euler was expected to take
sides. Wearily, he chose the path of least resistance—to support
what the Academy's president, Maupertuis, wanted. Maupertuis,
on the other hand, was an aggressive fighter who was seldom between feuds. He quarreled continually with Voltaire, members of
his own Academy, and other scientists, always dragging Euler
along with him. When Maupertuis finally became ill—"owing to
an excess of brandy," remarked Voltaire—and died, Euler was the
most likely candidate for the position of president. But Frederick
did not like Euler and was reluctant to appoint him. Euler was too
bourgeois, too obtrusively pious, too unsophisticated and unpol-
ished. Frederick cruelly ridiculed "that great cyclops of a geometer" and constantly tried to get the men whose intellects decorated
the court to admit that mathematics was really not very important.

Frederick's distaste for Euler may have sprung in part from the
fact that Frederick was never very good at mathematics and re-
sented someone who was better. Also, he preferred men like Voltaire—literary, witty, sophisticated men whom he could meet on
his own level to discuss "the immortality of the soul, freedom, and
Plato's hermaphrodites." He could show his poems to Voltaire—
Frederick's great ambition was to be a writer—and receive ef-
fusions of praise from his idol. By Euler he could only be told
that the sides of a triangle are proportional to the sines of the
opposite angles.

Frederick decided to bypass Euler and invite the French mathematician, D'Alembert, to be president. D'Alembert, however, unlike his benefactor, had a sense of propriety and justice. He refused the invitation, saying that it would be absurd to put anyone over Euler.

Meanwhile, Euler had kept up his relationship with the Russian academy. He continued to send them papers for their journal and entertained visiting Russian mathematicians and students. The esteem in which he was held by Russia—contrasting with his low status in Prussia—was illustrated during the 1760 invasion of Berlin by Russia. The invaders assigned two men to protect Euler's town house from looters. Unfortunately, his country house was inadvertently destroyed by the invading army; but when the Tsarina heard about it, she immediately sent 4,000 crowns indemnity.

The difference in treatment was not lost on Euler. He finally decided to leave Berlin, his home for the past twenty-four years, and to return to Russia.

When he arrived in St. Petersburg, the grateful Catherine the Great presented him with a completely furnished house, her own royal cook, and saw to it that his sons were appointed to good positions. Although eight of Euler's thirteen children had died, his household now comprised eighteen people counting in-laws and grandchildren.

Shortly after his return to Russia, Euler developed a cataract in his remaining eye and went completely blind. Now, the average blind mathematician is about as useful as a blind painter—but Euler was not average. He continued his work—increasing it if anything, and in the next few years submitted over four hundred treatises to the Russian Academy. He composed by dictating to a secretary who knew next to nothing about mathematics (he was a tailor by trade) and by writing formulas in chalk on a large slate. The secretary merely copied them down in the manuscript.

Euler's remarkable memory stood him in good stead during his years of blindness. He calculated long and difficult problems in his head—sometimes to as many as fifty places. Equally as remarkable, he completely memorized the *Aeneid* and could recite the whole thing, word for word, noting where each page ended and the next began.

In 1771, another disaster struck. His house caught fire and the poor blind man was trapped, unable to escape through the smoke and flames which he could feel but not see. A servant, Peter Grimm, dashed into his room and carried him out to safety. Euler's manuscripts were also saved, but everything else burned.

Later the same year his bad luck took a temporary change for the better. He had a successful operation to remove the cataract and after five years of total darkness could see again. But a few weeks later, infection set in, accompanied by days of almost unbearable agony. When it was over, Euler was once again in total darkness.

The faith that the sophisticated Berliners had ridiculed and a strong innate optimism carried Euler through these many trials. When his wife died in 1776, the indomitable mathematician courted her half-sister and married her a few months later. He was sixty-nine at the time.

It is said that Euler calculated as long as he breathed. Old—past seventy—blind, and slightly deaf, he continued to produce his prodigious works. At Catherine the Great's request, he turned his hand to writing a book on elementary algebra—nothing was beneath him and his books for beginners show the same superb organization and elegant style as his more advanced writings.

Euler loved mathematics and rejoiced over each new discovery, whether it was his own or not. He never seemed to mind if his theories were criticized or exploded nor indulged in quarrels over priority of discovery—it was the mathematics and not his reputation that was important. When Euler had been in Berlin, a twenty-three-year-old boy sent him a method of solving a certain type of problem in calculus—a method Euler himself had only recently discovered. Euler sent his compliments and encouragement to the young man, informing him that he was delaying his own publication of the method "so as not to deprive you of any part of the glory which is your due." The young man, Joseph-Louis Lagrange, went on to become one of Europe's most eminent mathematicians. Euler's generosity in helping Lagrange become recognized is unmatched in a field where jealous feuding, backbiting, and selfish credit-grabbing have been the standard practice of many of its greatest men.

One interesting problem Euler posed the year before he died involves what are known as Greco-Latin squares. The problem arose when he was asked, after others had failed, to arrange six soldiers of six different ranks from six of Catherine II's regiments. The men were to be lined up in six columns, six men to each column, so that in each row and column there would be a man from each regiment as well as a man of each rank. After shuffling through many different arrangements, Euler concluded that it was impossible and claimed that it is also impossible to do the same thing if ten different ranks and regiments are represented or any number of ranks and regiments that is not an even number divisible by four. That is, it would be possible with eight or twelve rows and columns, but not six, ten, or fourteen. This was merely conjecture on Euler's part. He could not prove it nor test each case by trial and error. Even modern computers working more than a hundred hours can test only a fraction of the possible combinations for ten ranks and ten regiments. It would take at least a century to test all possible cases of ten ranks.

For 177 years, mathematicians went along with Euler's conjecture, and then in 1959, it was announced that a group of Americans, dubbed "Euler's spoilers," had disproved it by arranging a Greco-Latin square of ten orders. Euler, in his Calvinist Heaven, is probably rejoicing as much as the Americans.

The problem in Euler's words "had little use in itself," but like so much that is thought to have no practical value has shown itself to be very useful. Scientists can use the solution in testing large blocks of different subjects with a minimum of duplication. For instance, they can test ten different drugs in ten different strengths on ten different national groups, each group consisting of ten subjects ranging in age from one to ten. By using Greco-Latin squares, this test can be performed at one time without any duplication, as shown in Diagram 24. The ten rows represent the ten different national groups. The ten columns represent the ten age groups. The digits from 0 to 9 represent the ten drugs and the letters represent the ten strengths. Each drug in each

Age Groups

0A	4H	1I	7G	2J	9D	8F	3E	6B	5C
8G	1B	5H	2I	7A	3J	9E	4F	0C	6D
9F	8A	2C	6H	3I	7B	4J	5G	1D	0E
5J	9G	8B	3D	0H	4I	7C	6A	2E	1F
7D	6J	9A	8C	4E	1H	5I	0B	3F	2G
6I	7E	0J	9B	8D	5F	2H	1C	4G	3A
3H	0I	7F	1J	9C	8E	6G	2D	5A	4B
1E	2F	3G	4A	5B	6C	0D	7H	8I	9J
2B	3C	4D	5E	6F	0G	1A	8J	9H	7I
4C	5D	6E	0F	1G	2A	3B	9I	7J	8H

National Groups (left side label)

DIAGRAM 24

strength appears only once. Each age group and each national group is tested with all ten of these drugs in different strengths without any duplications or omissions. Simple as it seems, it took almost two centuries and countless hours of thought by men and machines to show how it could be done.

In his seventy-seventh year, on November 18, 1783, Euler was sitting at the table having tea with one of his grandchildren when he suffered a stroke. "I am dying," he cried, and minutes later became unconscious. He died a few hours afterward.

Carl Friedrich Gauss
1777-1855

Archimedes, Newton, and Gauss. These are the immortal triumvirate in mathematics. Archimedes stands as the door closes on Greece; Newton opens it to the modern age; and Gauss, a hundred years later, ushers in a new era of mathematical triumphs.

Gauss's father, Gebhard, was a part-time gardener and full-time laborer in the little town of Brunswick, Germany, where he toiled his life away the year round, just as his father and his father's father before him had. He took himself a wife, Dorothea Warnecke, who bore him one son, Johann Georg Heinrich, before she died at the age of thirty. Gebhard remarried, taking as his second wife an illiterate maid by the name of Dorothea Benze, who also bore him one son on an April day in 1777. He was christened Johann Friedrich Carl—and with him the cycle of back-breaking work and heritage of ignorance ended for that particular peasant family. From poverty and a humble background, Carl, as he was called, rose to the highest heights in mathematics' history.

Even as a toddler Carl showed signs of genius, which his parents interpreted as indicating an early death, for God's favorites die young. Carl could add and subtract almost before he could talk. One day while his father added up a long row of figures, three-year-old Carl watched patiently and when the sum was written down, exclaimed, "Father, the answer is wrong. It should be

———." Gebhard Gauss readded the figures and discovered that his son was right—there was an error and the answer Carl had given was the correct one.

The little prodigy learned to read as mysteriously and easily as he had learned to add. He implored his father to teach him the alphabet and, then, armed with this knowledge, went off and taught himself to read.

His precocious achievements were proudly displayed as though they were parlor tricks. Little Carl was popped into a chair and asked to add figures his father wrote on a slate while an audience of friends and relatives looked on admiringly. Unfortunately, Gauss inherited poor eyesight as well as genius and was unable to see the numbers. Too shy to admit it, he simply sat there while admiring looks turned to nods of "I thought so."

Parlor tricks are one thing, genius is another—and Carl's father was either unable or unwilling to recognize the latter in his son. He set him to spinning flax in the afternoons in order to supplement the family income, and had every expectation that Carl would learn a trade of some sort—perhaps weaving, like his uncle Johann Benze, whom Carl adored. It was Johann who first recognized and cultivated Carl's talents, evidently seeing in his nephew the hopes for all his own frustrated dreams.

At the age of seven, Carl was sent to the local grammar school, where the tyrant of a teacher thought nothing of using a whip to beat an education into the boys. To keep the class busy one day, he assigned them the problem of adding all the numbers from one through a hundred. When the pupils finished, they were supposed to lay their slates on the table in the front of the room. The teacher had no sooner stated the problem than Carl scribbled the answer on his slate and tossed it on the table saying, "*Ligget se*," low German for "There it is." No one had ever told Carl the formula for adding a sequence of numbers, and the teacher was astounded that he had discovered it for himself. It is the same formula that the Pythagoreans had used as a password in their secret society: $\frac{1}{2}n(n + 1) = S$, where S is the sum and n is the last number of the sequence 1, 2, 3, . . . n. Gauss probably figured out the solution by adding 100 and 1, 99 and 2, 98 and 3, and so on. In each case the answer is 101, and since there are one hundred numbers to be added, there are fifty sets of 101. Fifty times 101 is 5,050, the answer to the problem. (Or, by the formula, $\frac{1}{2} \times 100 \times 101 = 5,050$.)

Gradually it dawned on the teacher that Carl was something special—not simply bright, but more than that. As he himself could teach the child no more, he turned him over to his young assistant, Johann Martin Bartels, whose main duties were to make quill pens and teach the students how to use them. He also called in Carl's father to discuss the education of his brilliant boy. Apparently Gebhard Gauss left the interview with the first inkling that his son would not be a tradesman or laborer or even a small businessman. He would be something better—a doctor, a lawyer, or a professor. Legend says that Gebhard went home and threw out Carl's spinning wheel. It is more likely that the thrifty man kept the wheel but excused his son from using it.

From then on, Carl spent most of his time studying. But when it began to get dark, he had to stop and shuffle off to bed, for the Gausses could not afford to burn candles in the evenings. Carl overcame this obstacle by ingeniously fashioning his own lantern out of a hollowed-out turnip filled with fat, with a bit of cotton or an old rag for a wick. With this he could read late into the night.

News of Gauss's prodigious talents came, in fairy-tale fashion, to the ears of the Duke of Brunswick. Impressed by the rumors, he sent a servant to the humble Gauss cottage to bring Carl to the castle. The servant accosted Carl's half-brother, Georg, and believing him to be Carl, told him to come along, the Duke wanted to see him. Frightened by the sudden summons, Georg protested that there must be some mistake and persuaded the servant to take Carl instead. And so Carl trudged off to begin a friendship with the Duke that lasted until death. In later years, when Carl had become a world-famous mathematician, his brother Georg often commented that "I could be a professor now; it was offered to me first but I didn't want to go to the castle." Georg became a tailor and, after a stint in the army, settled down as a gardener.

At fifteen Carl was sent to college, with the Duke paying all expenses plus a modest allowance. Gauss studied ancient and modern languages—subjects his father considered of no earthly use—as well as mathematics. Three years later, when he entered the University of Göttingen, he was still undecided whether he should pursue mathematics or languages. His decision in favor of mathematics was finally made on March 30, 1796, the day he discovered how to construct a 17-sided polygon using only a compass and straightedge. Even as an old man, with dozens of important works behind him, he considered this construction as one of his greatest discoveries.

On the same March day that he made his decision, he began keeping a sort of mathematical diary in which he recorded the scores of mathematical ideas and problems that stormed his mind every day. The creative floodgates were opened to a veritable deluge of ideas. He had already stumbled across the possibility of a non-Euclidean geometry; was making tremendous strides in arithmetic, or number theory; and was in the process of discovering a graphic representation of complex numbers and a proof of the fundamental theorem of algebra. Gauss's youth, like Newton's, was an extremely prolific and creative period. His mind, he said, was bombarded by so many ideas that he could record only a small part of them and of these had time to develop only a fraction.

At Göttingen, Gauss was able to compare his own work with that of the masters—and the comparison was favorable. The more he read and compared, the more he realized that he himself was one of the mathematical greats. Even at the age of twenty-one he had already done work that would place his name near the top.

Yet, like his idol Newton, he kept his discoveries to himself and regretted that none of his friends was able to discuss the profound theories that interested him. One of his best friends at the University, Wolfgang Bolyai, a Hungarian noble whose son later made a mark for himself by discovering one kind of non-Euclidean geometry, remarked that Gauss "was unpretentious and did not make much of a show . . . One could be around him for years without ever suspecting his greatness."

Yet Bolyai himself recognized and admired Carl's genius. One year he and Gauss hiked from Göttingen to Brunswick, where they visited with Carl's family. Frau Gauss, wanting to be assured that all her son's studies were not in vain, managed to corner Bolyai alone and ask him if it were true that Carl was as brilliant as everyone said, and if so, what would he amount to. Bolyai answered, "The first mathematician in Europe," whereupon Dorothea Gauss burst into tears.

At twenty-one, Carl said good-by to his friends at the University, many of whom he would never see again, and in Bolyai's words, "led by angels of the temple of fame and glory," returned to Brunswick. For some reason, possibly the presence of his father, he did not live at home. Gauss never made any secret of the fact that he did not like his father, whom he termed "domineering, uncouth, and unrefined." To return to his father's badgering questions of what he was going to do with his fancy education would have been difficult, for Gauss himself did not know. Any plans he did have evidently fell through, for he wrote to his adoring friend, Bolyai, "my financial prospects are all shattered." He had no money and desperately needed some, for he was living only through the kindness of creditors. The Duke of Brunswick soon assured him, however, that his allowance would continue and not to bother his mind about money matters.

Gauss spent the first few months traveling back and forth between Brunswick and Helmstedt, where he went to use the library. He was absorbed in finishing a manuscript on the theory of numbers that he had begun at Göttingen. Entitled *Disquisitiones Arithmeticae,* the book was held up in printing for three years and did not appear until 1801. With it, Gauss's reputation was made. To discuss its contents here would take us too far afield, but a simple and exceptionally good explanation of its major subject—the theory of congruences—can be found in *Foundations of Mathematics* by Denbow and Goedicke (Harper & Brothers), pages 534-544.

Gauss apparently planned to use the *Disquisitiones* as his doctoral dissertation, but when the printing of it bogged down, he wrote a shorter but equally important piece, imposingly entitled *Demonstratio nova theorematis omnem functionem algebraicam rationalem integram unius variabilis in factores reales primi vel secundi gradus revolvi posse,* which was published in 1799.

Briefly, the title means that Gauss had found a proof for the fundamental theorem of algebra: Every integral rational equation in a single variable has at least one root. Stated more simply, the theorem means that any algebraic equation has at least one root. The corollary is even more important: Every equation will have as many roots as the highest power of the unknown. That is, $x^4 + 2x^3 + 9 = 0$ will have four roots; $x^3 + x^2 + 2x + 4 = 0$ will have three, and so on. In some cases, some or all of the roots will be identical, yet each must be counted as a separate entity. The reason for this is that, as was shown at the beginning of the seventeenth century, if $x - a$ is a factor of an equation, then a is one root of the equation. This relationship holds true in any equation. For example, $x^2 - 6x + 9 = 0$ can be factored to $(x - 3)(x - 3) = 0$. Its roots are then 3 and 3, which, although identical, are each counted separately.

By proving the fundamental theorem, Gauss took one more step toward systematizing algebra and generalizing its rules.

In asserting that every algebraic equation has at least one root, he had to make sure that the number system was complete. If he omitted irrational numbers, for instance, equations having only irrational roots would be insoluble and it would not be true that every equation has at least one root.

Is the number system complete with the negative, positive, rational, and irrational numbers which make up what is called the real-number system? The answer is no. A simple equation such

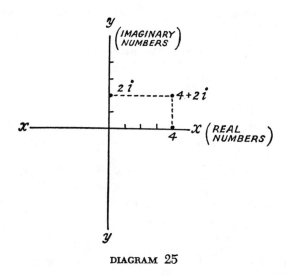

DIAGRAM 25

as $x^2 + 4 = 0$ cannot be solved if only real numbers are used, for the answers would be $x = \pm\sqrt{-4}$, or $x = \pm 2\sqrt{-1}$. And as Euler had said, ". . . such expressions as $\sqrt{-1}$ and $\sqrt{-2}$ are impossible or imaginary numbers, since they represent roots of negative quantities; and of such numbers we may truly say that they are neither nothing, nor greater than nothing, nor less than nothing, which necessarily makes them imaginary or impossible." Gauss, however, believed that "just as objective an existence can be assigned to imaginary as to negative quantities," and admitted them to the number system.

Not only did he admit them, he put them on a sound basis as well by showing that imaginary numbers, like real numbers, can be plotted on a graph. In Diagram 25, the x axis is used for real numbers and the y axis for imaginary numbers. To plot $2\sqrt{-1}$ simply mark off two units along the y axis (Diagram 25). (Gauss also originated the use of "i" to represent $\sqrt{-1}$; thus $2\sqrt{-1}$ is written as $2i$, $\sqrt{-2}$ is written $i\sqrt{2}$, etc.) A combination of imaginary and real numbers, such as $4 + 2i$, can also be plotted. In Diagram 25, the real and imaginary numbers determine a point, which Gauss considered to be a separate number called a complex number. From these complex numbers all other numbers are derived. That is, any real number is actually a complex number of the form $a + bi$, where a is real and b is zero; any imaginary number is a complex number of the same form with a zero and b real. The number system, domain, or field—whichever term is preferred—can now be diagramed in this way:

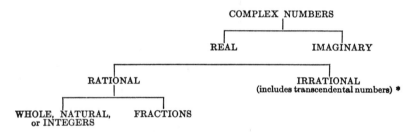

Of course, any of these numbers can be negative or positive.

The admission of complex numbers not only affected algebra, but analysis and geometry as well. The theory of functions of a complex variable developed; absolute differential geometry and vector analysis—so vital to modern science—grew out of these half-real, half-imaginary numbers.

Complex numbers can be added, subtracted, multiplied, divided, raised to a power, or have their roots taken and in every case a complex number of the form $a + bi$ results—although either a, b, or both may be zero. No new kind of number has to be invented in order to operate on complex numbers. The alge-

* Transcendental numbers, which are nonalgebraic, will be discussed later.

braic number system is now closed, for with the admission of complex numbers, any algebraic equation can be solved and all operations performed. This achievement of closure in mathematics is the goal for which men had been searching since the days of Pythagoras. It is a search that has parallels in other fields of science. Chemists strive to find all the elements, chemistry's counterpart to numbers. Over one hundred have already been discovered and identified. Physicists, in turn, want to know the answer to the even more fundamental question: what are the basic units of matter. At first they believed that atoms were the indivisible units—the very word means indivisible—but then atoms were found to be made up of electrons, protons, and neutrons. Yet even this does not complete the list. New particles such as antiprotons, positrons, and neutrinos are constantly being added— and who knows whether all these cannot be added to or broken down even further. All these goals correspond to the mathematical search for closure which was finally achieved by Gauss.

One minor discovery of personal importance to Gauss might be mentioned. His mother had not known the date of his birth except that it was eight days before Ascension. In 1800, Gauss formulated the rule for determining the date of Easter, which enabled him to place his birthday on April 30.

Gauss received his doctorate from the University of Helmstedt for his dissertation, and, still supported by the Duke of Brunswick, spent most of his time getting *Disquisitiones Arithmeticae* ready for the printer.

At some point during this period, it is thought that he either accepted or contemplated taking a teaching position and with this end in mind, wrote a short essay entitled *The Metaphysics of Mathematics,* which is one of the simplest and clearest discussions ever written on the foundations of mathematics. It may seem that mathematics is a difficult enough subject without getting into the metaphysics of it—and yet this is quite often, in Gauss's hands, the easiest part. Indeed, it is the very simplicity of his essay that has led to the belief that it was intended for beginning students.

In this essay, Gauss aimed at cutting away all the complicated, abstruse operations and theorems until he was left with only the essentials, the core or foundation of mathematics. Mathematical discoveries and applications had been proliferating at a fantastic rate since the time of Newton, like an untended garden with flowers and vines and weeds rambling in such profusion that it was impossible to see any pattern in their growth or to examine the soil from which they sprang. And it was the soil and roots that interested Gauss. As a student he had often confided in Bolyai his belief that the foundations of mathematics needed reworking and investigating; the simple fundamentals of arithmetic and geometry which had been accepted so long ago and so completely that men now took them for granted, Gauss eyed with suspicion.

His overpowering critical analysis of the most obvious mathematical "truths" was one of the characteristics—and perhaps source—of his mathematical genius. As early as 1792, when he was only fifteen years old, he was looking askance at Euclid's definition of a plane and axiom on parallel lines. His researches on these alone would have been enough to make him famous, but Gauss never published his remarkable findings. The honors here were reaped by other men. It is known for certain, however, that he succeeded in creating another kind of geometry—non-Euclidean geometry—by eliminating a dubious axiom and the conventional flat plane. To have found his suspicions of these two basic "truths" justified must certainly have inspired him to look more closely at other "self-evident" truths.

In his essay, *The Metaphysics of Mathematics,* he announces no astounding discoveries, but does bring order out of chaos. Mathematics he defined as the subject that deals with relationships between magnitudes and then illustrated how these relationships are determined or constructed. He divided magnitudes into two groups: the first consisting of extensive or spacial magnitudes, such as lines, solids, and angles; the second, of intensive magnitudes, such as velocity and density, which are all dependent on the extensive magnitudes. That is, velocity as a magnitude is dependent on space; it is measured by the time taken to get from one point in space to another point.

It is impossible to study a magnitude by itself, noted Gauss. A line in itself offers nothing to investigate. But put another line next to it and certain observations can be made. Lengths and directions can be compared; angles can be measured, and other relationships studied. Thus, mathematics considers how magnitudes relate to each other.

Arithmetic studies only the relationships of magnitudes to each other; geometry studies this, too, and also the relationship of the magnitudes' *positions* to each other. The ancient Greeks preferred the purely geometric method; under the influence of the Arabs in the Middle Ages, this method was discarded in favor of the arithmetical one. Even geometrical relationships that are purely positional began to be studied arithmetically. Thus trigonometry describes positional relationships in arithmetic terms. The modern preference for the arithmetical method over the geometric one, Gauss quite rightly attributes to our number system—which the Greeks did not have. With an efficient number system, men quickly discovered the superiority of the arithmetical method to the cumbersome geometric one.

In studying the different relationships between magnitudes, mathematics has made a basic distinction between those relationships that involve the use of infinity and those that do not. The former relationships are considered to belong to higher mathematics; the latter to lower, and within each of these fields even further distinctions are made.

Several kinds of relationships can exist between magnitudes, the most basic being that of the whole to its parts. Addition consists of finding the whole from its parts; subtraction, of finding one part from the whole and another part. All other operations are adaptations of addition and subtraction.

Such was the nature of Gauss's imposingly titled essay, *The Metaphysics of Mathematics,* whose almost ridiculously obvious descriptions of mathematics presaged a whole movement in the investigation of the bases of mathematics, a movement that in the end discarded much of Gauss's obvious description as being too complicated, too imprecise, and not at all obvious! The best-known and perhaps most ambitious attempt to weed out the extraneous and reduce mathematics to a handful of seeds—and then even to analyze the basic properties of the seeds themselves—was made by Bertrand Russell and Alfred North Whitehead in their formidable and largely unread *Principia Mathematica.* Their object was to reduce mathematics to a system of a few basic axioms which, when combined in various ways, would give rise to the whole mathematical structure, just as the twenty-six letters of the alphabet are combined in various ways to produce literature. To do this they had to rid mathematics of all "self-evident" assumptions whose actual bases were empirical. Numbers, for instance, had developed from man's efforts to enumerate objects and therefore could not be admitted to the system unless they were put on a sounder basis, a purely axiomatic basis. Euclid's axioms, next to theirs, are complex constructions bristling with hidden and unstated assumptions. Yet, despite the simplicity of the Russell-Whitehead axioms, there is doubt that they are, indeed, the basic building blocks of mathematics or that such fundamental axioms can even be found.

The whole microscopic inspection of mathematics was inspired, as shall be seen later, by the flaws in Euclid's "self-evident" axioms. Gauss, having privately observed these flaws for himself several years earlier, was therefore one of the first to examine the mathematical structure with caution and suspicion. He questioned even such obvious truisms as the existence of three-dimensional space. "We can imagine," he said, "beings who recognize only two dimensions; beings of a higher state of existence might have a similar contempt for us and our three dimensions." Just because the human mind can conceive of only three dimensions does not mean that a larger number cannot be. Thus, it is possible that the reality man sees is only a very limited picture of what actually exists, a picture limited by the human mind and senses. Reality may be not only more than meets the eye, but more than can be imagined by the human mind. And yet, as was later shown by Cayley, this same human mind can handle through mathematical symbols the material of that inconceivable reality! Limited, earthbound, the mind can yet fly. Imprisoned in a three-dimensional world it can yet free itself through mathematics to explore a four- or five- or n-dimensional structure.

Nor has mathematics stopped with inconceivable geometries. The whole trend in twentieth-century science is away from depicting phenomena visually or even pretending that such a visualization is possible. Nothing but equations are used to describe most phenomena on atomic and macrocosmic levels. This modern trend away from pictorial representation of physical activity is in part due to the impossibility of such visualization of many phenomena and in part due to the suspicion that such visualization inevitably leads to errors. Euclid's diagrams, for instance, served not only to clarify geometry, but to camouflage a multitude of its unstated, unwarranted assumptions.

In non-Euclidean geometry, Gauss had discovered what is probably one of the most startling and revolutionary concepts in the history of human thought—and he buried it away in his notebook where it was unearthed only after his death. His regrettable reticence on the subject was perhaps prompted by the fear of censure. Gauss, like Newton, shied away from controversy, and controversy was what he could expect if he published his findings.

Equally as regrettable, but more excusable, was his next step. At the age of 24, he dropped his researches in pure mathematics and turned to astronomy. This misuse of genius was not entirely his fault. Unable to get a university appointment commensurate with his abilities, plagued by financial worries—he could not expect the Duke of Brunswick to support him forever—he chose the fastest road to academic recognition and fame and fortune: astronomy. Several small planets, or planetoids, had recently been discovered and Gauss put his mathematics to work by calculating their orbits. He announced his first findings in 1801 and was immediately hailed for his work. The St. Petersburg Academy of Sciences elected him to membership and its observatory offered him the position of director. The offer was counterbalanced by the Duke of Brunswick who raised Gauss's allowance and promised to build him an observatory in Brunswick. Out of loyalty to his patron, Gauss rejected the Russian offer. Europe's greatest mathematicians all looked to the young Gauss as their equal and friendly efforts were now made to secure him a position at the University of Göttingen. The negotiations dragged on for five years before a definite agreement was reached. Meanwhile, Gauss continued his researches in astronomy at Brunswick.

During this period, he met "a splendid girl" by the name of Johanna Osthoff, the daughter of a prosperous tanner. Gauss was twenty-six when he first met her and almost immediately decided that she was "exactly the kind of girl I have always wanted as a life companion." He courted her for two years before he managed to summon up enough courage to propose—and then did so by letter. Johanna, having heard a rumor to the effect that Gauss was engaged to someone else, ignored the proposal for three months. When she finally accepted, the lovesick mathematician sent off an ecstatic note to a friend telling of his betrothal to "an angel who is almost too saintly for this world."

They were married on October 9, 1805, and a little less than a year later, a child, Joseph, was born. Gauss was blissfully happy with his wife and infant son. His work was going well and his reputation, as noted previously, had already leapt the borders of Germany, spreading as far as Russia.

But all around this blissful existence was chaos. Napoleon was overrunning Europe. Germany was a conglomeration of small states which Napoleon easily swallowed, to form the Confederation of the Rhine. Only the largest and most powerful German state—Prussia—was still free. In 1806 its king, Frederick William III, threw his army against Napoleon. The Duke of Brunswick was made commander-in-chief, subject to the meddlesome interference of Frederick. Fourteen years earlier the Duke had opposed the same forces that Napoleon had ridden to power. He had led the armies of Austria and Prussia into France to save the throne from annihilation by the revolutionists. The Duke had been unsuccessful then. He was to be unsuccessful again. When his troops met Napoleon's Grande Armée near Jena, twenty thousand Prussians were taken prisoner and the Duke was mortally wounded by a musket ball; what was left of the army fled. The defeat was made even more humiliating by Napolen's refusal to allow the dying Duke to return home.

Gauss was standing at the window, looking out at the main highway, when a wagon hurried past carrying the injured Duke, taking him not home but to where he might at least die in freedom. Gauss watched as the man rode by who had been his friend, his patron, his advisor, more a father to him than his real father. On November 10, 1806, the Duke died, and a quiet, almost unexpressed but deep grief settled on Gauss. He became more reserved, more serious and developed a lifelong horror of violence and death.

A year later Gauss took his wife and child to Göttingen where he had just been appointed director of the Observatory. It is one of the ironies of history that Napoleon later spared the city because "the foremost mathematician of all time lives there," yet had shown no such mercy toward the Duke of Brunswick, who had been responsible for educating that "foremost mathematician."

The death of the Duke marked the beginning of a period of

almost unbelievable tragedy in Gauss's life. Within the next
three years his father; his dear uncle, Johann Benze; his wife;
and his third and youngest child all died. In addition, he was
overwhelmed with financial difficulties, for his salary was low
and his stipend from the Duke had, of course, stopped. Friends
and relatives criticized him for wasting his time in research when
he could make more money elsewhere. Gauss became so depressed that he confided to his diary: "Death is dearer to me
than such a life."

The most painful blow was undoubtedly the death of his wife
on October 11, 1809. "Last night," he wrote the day after she
died, "I closed the angelic eyes in which I have found Heaven for
the last five years." The great emptiness could never be filled, yet
Gauss could not bear to "wander in loneliness among the happy
people who are all around," and six months later offered his "divided heart" to Friederica Wilhelmine Waldeck, daughter of a
Göttingen professor. She accepted and they were married on
August 4, 1810, less than a year after Johanna's death.

In the next six years, three children were born to them—
Eugene, Wilhelm, and Therese.* Then Minna, as Gauss's wife
was nicknamed, contracted tuberculosis. Gauss nursed his ailing
wife, cared for the children, and sent for his widowed mother to
come live with them. Neither wife nor mother could have asked
for a more dutiful husband or son. Although he had to be away a
great deal on a geodetic survey, he wrote every day, sometimes
"at midnight with my eyes almost closing."

His work in geodesy lasted approximately ten years, from 1820
to 1830. It is impossible to describe all the mathematical conse-
quences of this work. Differential geometry, theories of surfaces,
statistics, the theory of probability—all were developed by Gauss
and others as offshoots of this geodetic work. For instance, his
contributions to the theory of probability came about through the

---

* Their full names were Peter Samuel Marius Eugenius, Wilhelm August
Carl Malthais, and Henrietta Wilhelmine Caroline Therese.

thousands of sightings he had to make in his survey. Each sighting or reading was made several times, often with different results, for the human eye is not infallible. By applying his method of "least squares" which he had invented as a teenager Gauss was able to determine which of the readings was most probably correct. Furthermore, by plotting the results of the readings, Gauss

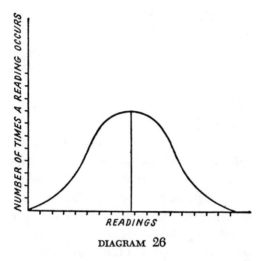

DIAGRAM 26

showed that the graph always results in a bell-shaped curve (Diagram 26), often called the Gaussian curve, with half of the errors to the left of the center and half to the right. In other words, the distribution of errors always follows a pattern which is described in the Gaussian law of normal distribution of errors. Thus, even "accidental" errors can hardly be called accidents, for according to Gauss, they *must* occur with a frequency that can be mathematically estimated *in advance*. The larger the sampling, the more accurate the estimate of the number of errors will be.

The Gaussian law of normal distribution is obviously of great value to anyone trying to determine what is the correct measurement from a mass of data. But the application of the law does not stop there. Scientists soon found that not only errors but all sorts of other phenomena obey the same law. The heights of men, their intelligence and other abilities all conform to the same bell-shaped curve. The incidence of the idiot and the genius, the pygmy and the giant are as mathematically determined as is the Sun's course in the skies.

The connection between a survey of the contours of the earth and a study of the number of imbeciles at first seems nonexistent. But the fact that the connection does exist only testifies to the tremendous scope of applied mathematics.

After the geodetic project was completed, Gauss returned home and, except for one night, never again slept anywhere but under his own roof. He had had enough of the inconveniences of traveling and sleeping in strange beds and getting up before dawn to take readings in the next town.

Meanwhile, Minna's health had deteriorated to the point where she was completely bedridden. Gauss spent his days nursing her and working on his mathematics—both of which he did so intensely that he thought of nothing else. The day that Minna died, September 12, 1831, so a story goes, Gauss was in deep concentration on a problem when a servant rushed in to tell him his wife was dying. So absorbed was Gauss in his work that although he heard the words, he failed to comprehend their meaning. "Tell her to wait until I've finished here," he muttered.

Whether the story is true or not is unimportant, but it does illustrate that like his two peers, Newton and Archimedes, Gauss was capable of intense concentration, and obsessed and totally absorbed by mathematics. He lived and breathed it; he was overwhelmed by it, more ideas crowding his mind than he could ever hope to handle. And he gave himself up to the subject completely; everything became the object of mathematical inquiry. In a notebook he kept a record of a mass of seemingly trivial information: the number of steps from his house to the observatory and library; the number of days friends and famous men had lived; the ages of his children when they took their first steps, cut their first teeth, were vaccinated. Every insignificant mathematical fact was noted and stored up for future use. His lists of the number of steps from one place to another flowered into a topology study. The records of the ages of people formed the basis of an actuarial study in connection with pensions. Without even trying he learned most logarithms by heart. His remarkable memory for figures enabled him to calculate large numbers mentally, and once he had made the calculation, the answer stuck in his mind for years. His son, Eugene, apparently inherited this ability to handle numbers mentally and in his old age amused himself by calculating long numbers in his head. That he never became a mathematician was due to his father's influence. Gauss dissuaded his sons from mathematics—some say because he was afraid his boys would detract from the Gauss name; the more generous say because he knew that his sons would surely lose in the inevitable comparisons between them and himself.

An acquaintance once described Gauss as a megalomaniac suffering from "paranoia geometrica." This observation, while harsh, is true. Gauss's absorption in mathematics was so complete as to appear slightly mad—especially to the nonmathematician. In this respect, however, Gauss is no different from all great artists who even in moments of leisure are collecting material for the next book, painting, or symphony. He is no different from the successful—or even unsuccessful—business tycoon who talks and thinks business on the golf course, over lunch, and on the train. It is generally sad but true that success in almost any field demands a constant and single-minded dedication, with no time out to explore other areas. And so it was with Gauss. Mathematics was everything—his work, his pleasure, his passion. "I have been living here," he wrote as a young man, "chiefly for my goddesses—the sciences." And mathematics was to him the queen of all the sciences, a work of art. "You have no idea," he once exclaimed to his students, "how much poetry there is in the calculation of a table of logarithms!"

He was committed to perfection in his field—perfection, that elusive ideal that is the never-realized goal of all art. And mathematics *is* art—perhaps the most sublime of all arts. Bertrand Russell has described it as possessing "not only truth, but supreme beauty—a beauty cold and austere, like that of sculpture, without appeal to any part of our weaker nature, without the gorgeous trappings of painting or music, yet sublimely pure, and capable of a stern perfection such as only the greatest art can show." * It was this perfection that Gauss strove for and instilled in his pupils. His demands for more rigor and less vigor have provided the direction for all of modern mathematics. The old problems that had plagued mathematicians from the days of the Greeks—irrational numbers, infinity, and infinitesimals—were resurrected and attempts made to solve them. Cantor, Cauchy, Weierstrass, and Dedekind—the mathematical heirs of Gauss—explored these neglected fields and shed new light on the basic concepts of numbers, infinity, infinitesimals and limits, etc.

Gauss's passion for perfection influenced his method as well as his subject matter. His writings are models of perfection. Theorems are proved rigorously and elegantly, with no unnecessary details to crowd the picture and no necessary steps omitted. In spite of this—or more likely because of it—his books are considered to be very difficult, whereas Euler's leisurely, imaginative, and certainly unrigorous writings are considered to be the height of clarity. This paradox is not as strange as it seems. The mind—even a mathematician's mind—does not work step by step. It leaps ahead, backtracks, and leaps again. When confronted with the artificial mental processes of strict logic, it seems to stumble and become entangled in the close chains of reasoning.

Gauss never published anything until it was as finished and as perfect as he could make it—quite unlike his idol, Newton, who in his prime confidently appended problems and undeveloped

* Bertrand Russell, *Mysticism and Logic*, published by George Allen & Unwin, Ltd.

ideas to his books so that others could work on them. Not so with Gauss. "I get no pleasure from incomplete solutions," he explained, "and work in which I have no joy is torture to me." In keeping with this practice, he adopted a seal depicting a tree with a few large apples and the motto *"Pauca sed matura"* and always upheld it. As Gauss got older, his reluctance to publish increased, and more than one mathematician remarked on his selfishness in keeping everything to himself and never encouraging or helping younger mathematicians. Furthermore, Gauss's habit of brushing aside the discoveries of others by announcing that he had discovered the same thing years before but had not bothered to publish it, did nothing to raise his popularity.

Yet in his lifetime, he did announce the results of a tremendous amount of work in both theoretical and applied mathematics. Algebra, geometry, analysis, and arithmetic or number theory—all four branches were further developed by him, especially the last. He often said that mathematics was the queen of sciences and arithmetic was the queen of mathematics. In this royal circle, Gauss himself was honored with the title "Prince of Mathematics." He was able to grasp the whole immense field of mathematics— not just a portion of it—and was master of all he surveyed. When the French mathematician, Laplace, was asked who was the greatest mathematician in Germany he answered, "Pfaff." The astonished questioner exclaimed, "Not Gauss?" "Oh," Laplace replied, "Gauss is the greatest mathematician in the world."

In addition to his theoretical work, Gauss devoted many years to applied mathematics—and here he seemed to be a Jack-of-all-trades, working in astronomy, magnetism, topology, crystallography, optics, and electricity. In 1833, he demonstrated the possibility of sending telegraphic signals—thus anticipating Samuel Morse by at least three years.

Mathematics bestowed on Gauss all the rewards of success: fame, position, the joys of accomplishment, and the respect of his colleagues and students. In turn, this queen demanded of her prince a lifetime of constant work and concentration that led him to maintain that if others thought about mathematics as much as he did, they, too, could come up with the same discoveries and solutions. (Whether this was intended as a compliment on general intelligence or a criticism of the average man's sloth is not known.) The struggles—often lasting for months or even years—over problems that defied him, he once described to a friend: "All pondering, all searching was in vain. With regret I had to lay down my pen time after time. Finally I succeeded several days ago—not as a result of my tedious searching, but, you might say, by the Grace of God. The problem was solved like lightning—I myself could not even follow the train of thought. Strangely enough, the solution now appears easier than many others that did not take as many days as this had years."

This sudden insight into a problem is something almost every great mathematician possessed—Archimedes had it, Newton, Descartes and Pascal had it, as well as many others. Yet flashes of inspiration that come like thunderbolts from the blue are almost always preceded by days, weeks, or months of fruitless and intensive thought. Gauss's solution to the inscription in a circle of a regular 17-sided polygon, for instance, came to him one morning in just such a flash after he had brooded about it for weeks.

Gauss's professional successes were counterbalanced by personal tragedies and failures. Indeed, his devotion to mathematics served two ends: the positive one of professional accomplishment and the negative one of helping him forget his troubles. Immersed in "quiet contemplation," he admitted he was able to forget for a few hours the "unpleasant material world." And he had much to forget. Besides the untimely deaths of his wives, his favorite student, Eisenstein, whom he considered "one of the greatest geniuses of all time" and ranked with Archimedes and Newton, died tragically young. In 1837 his older daughter was banished from Göttingen because of her husband's political activities. Along with her, Gauss's close friend, Wilhelm Weber, and the remarkable Grimm brothers were expelled. Gauss never saw his daughter again, for she died three years later, still in exile.

The shock of seeing so many loved ones die or leave intensified Gauss's tendencies toward pessimism and introversion and colored all his views. His dread of death turned him into a morbid hypochondriac with a suspicious distrust of doctors; his hatred of violence, into a political reactionary who condemned the whole democratic movement sweeping through Europe, for he knew that reforms would have to be won with blood. His son-in-law's activities and resulting exile were to Gauss the height of folly—shadowed over by the memory of a beloved Duke who had died in another futile war thirty years before. Peace and quiet were his political and professional touchstones; he was opposed to anything that might upset this tranquillity. Revolution, of course, was anathema, whatever its ends might be. He wanted neither a strong leader nor a democratic system—the former led to tyranny, the latter to mismanagement. His cynical views leave one wondering just what kind of government he would have preferred.

Even his tastes in reading were affected by this morbidity. He loved literature and read the masters in their original languages—but avoided anything that did not have a happy ending. His preoccupation with death also spilled over into his work, where, as mentioned previously, he kept records of the number of days friends and great men had lived.

As his life wore on, his pessimism deepened. He confided his unhappiness to his old university friend, Bolyai, saying, "It is true that in my life I have won much that the world honors. But believe me, my dear friend, tragedy has woven itself through my life like a red ribbon . . . and the griefs overbalance the joys a hundredfold."

Only in his work could he find happiness. Close personal relationships were, for him, almost impossible. Like Newton, he was a genius, but neither a lovable nor a loving one. Except for one or two friends, his mother, wives, and daughters, Gauss had little affection for people, either as individuals or as a group. Colleagues and acquaintances were referred to with such epithets as "windbag" and "Dummkopf," and the world was populated by "stupid and immoral masses." His interest in his own brother was so limited that a year after Georg married Carl knew nothing about his wife—not even her name.

Disappointments, misunderstandings, and quarrels marred his relationships with his sons. Tender and devoted to the women in his family, he was domineering and abrupt with his sons. He pushed his favorite, Eugene, into law rather than encourage him in mathematical or scientific work, for which he showed an unusual talent. The boy became rebellious and irresponsible, and after a gambling spree, presented his irate father with a bill for his losses. A violent quarrel ensued, which was to be the last of many, for Eugene packed up and left home without even saying good-by. He went to America where he eventually became a successful banker. Gauss never forgave him—especially since he had left at a time when Minna was seriously ill. Gauss never saw Eugene again.

After Minna's death, his children left home one by one until there was only Therese left. Wilhelm joined his brother in America and began amassing a fortune that eventually totaled over a million dollars.

Joseph, Gauss's oldest child, married and left home in 1840. He is the only son who remained in Germany and the only one who did not cause his father heartbreak.

Gauss's mother died in 1839 at the age of ninety-seven, and on her Gauss had showered the devotion he never gave his sons. He cared for her tenderly during her terminal illness and through her years of blindness, a malady that seemed to be hereditary in the Gauss family. Her life had been none too happy—Gauss termed it "full of thorns"—and he did everything possible to bring her some measure of happiness. Her greatest joy undoubtedly came from seeing her son famous and respected. For herself she never asked anything but to wear her simple peasant dress, to eat with the servants in the kitchen, and to live the plain sort of life she always had.

After her death, Gauss was left alone at home with his youngest daughter, Therese, to care for the house. Since his tastes, like his mother's, were simple, his pleasures few and his social life practically nil, Therese's duties were minimal. Gauss was a thrifty man with no taste for luxury. His tiny study was furnished with a small table, a stand-up writing desk, and a single light. In his old age he added a chair and sofa. His bedroom was unheated and his food as simple as possible.

Gauss was now in his sixties, but his mind was as spry as ever. At the age of sixty-two he took up Russian as a sort of mental tonic and within two years was speaking it fluently. Besides languages (he knew French, Greek, Latin, English, Danish, and some Swedish, Italian, and Spanish), he also studied such far-afield subjects as bookkeeping and shorthand.

One concession to approaching old age was made: he gave up teaching and lecturing, which he had never liked anyway, considering them "a burdensome, ungratifying business" that was "a waste of my time." His reputation as a teacher—with a few notable exceptions—was poor. He refused to let his pupils take notes, insisting that they would learn more by paying attention instead, and had nothing but impatience for slow, or even average, students.

By the time Gauss neared seventy, his interest in the subject that had been his life's work began to wane. At the golden anniversary celebration of his doctorate, which was attended by mathematicians from all over Europe, one scholar noted that "it is no longer easy to have a scientific conversation with Gauss. He tries to avoid it and discusses only the dullest things in a continuous stream." This was the same man who had once called Gauss "that colossal genius."

What these dull things were, he did not say—perhaps politics or philosophy, which attracted Gauss at the end of his life, just as they had Newton. He spent his mornings in the library of the university where he gathered a large stack of newspapers from all over Europe—everything from the London *Times* to the local paper of Göttingen—piled them up on his chair, and pulled them out one by one in chronological order to read while students peered over their books to get a glimpse of the silver-haired genius of Göttingen.

His interest in philosophy and religion was undoubtedly heightened by the ever-present and inevitable prospect of death and the wonder of what came afterward. "Without immortality," he once said, "the world would be meaningless and all of creation an absurdity." Yet meaninglessness and absurdities do exist, and in this case, can be dispelled only by an act of faith. To believe or not to believe is not simply a matter of will. To a man whose mind is essentially scientific and analytical, faith is often alien. Gauss wanted to believe. "There are problems," he said, "to whose solutions I would attach an infinitely greater importance than to those of mathematics, for example, problems concerning ethics, our relation to God, or concerning our destiny and our future; but their solution lies wholly beyond us and completely outside the province of science." These are not the words of a devout believer, nor of an atheist. They are the words of a man who does not know, who is questioning. Gauss appears to have been an agnostic whose doubts left him stranded in a possibly meaningless, absurd world. To one of the few friends who still visited him in his old age, he confided these intimate fears and speculations. After one particularly long discussion he suddenly leaned toward his friend and with tears in his eyes asked, "Tell me, how does one begin to believe?"

Apparently even the advent of death could not bring faith, although it did bring a resurgence of real and imagined ills. He complained more and more of poor health, insomnia, and dyspepsia. At the age of seventy-seven, he began to suffer from "congestion in the chest," which was diagnosed as an enlarged heart, and was forced to get up at three every morning to drink seltzer water and warm milk for relief. His breath was so short that he could no longer walk to the library and could barely get about the house.

A year later, on February 23, 1855, after a losing fight against several heart attacks, he went peacefully "into that dark night" that he had dreaded for so long. Coincidentally, his pocket watch stopped ticking almost at the moment of his death.

He was buried at St. Albans Cemetery in Göttingen next to the unmarked grave of his mother. Like Newton, Gauss died a wealthy man although his salary had been modest. Money he had made from judicious investments was discovered hidden all over the house—in dresser drawers, desks, cabinets—enough to make a sizable fortune. But, also like Newton, his real legacy was his mathematics.

Nicholas Lobatchevsky
1793-1856

From the Greeks to modern Europe, mathematics steadily grew from a small sapling into a great tree with four major branches—geometry, algebra, analysis, and number theory. Smaller branches sprouted from the main limbs, and it seemed that the tree could grow indefinitely. Mathematical thinking had been grafted on to almost every other body of knowledge—in some cases successfully, in others with less promising results. Although the real might of mathematics lay in its applications to the physical sciences—chemistry, physics, etc.—the social sciences, too, adopted its methods. Politics, economics, sociology, and psychology all grew up under the tutelage of mathematics.

In the New World, a nation had been founded on principles that were mathematical in origin. The Declaration of Independence was inspired by the philosophy of John Locke who in turn got his ideas about political philosophy from mathematicians. What Newton had done for physical science, he attempted to do for social science. Starting with a few axioms or postulates, namely, that man is created free and equal with certain inalienable rights, Locke erected a whole political system. The Founding Fathers, in turn, adopted this philosophy as the basis of their new government.

In economics, Adam Smith borrowed freely from mathematical

thinking to produce a magnificently logical economic system, although as time went on, the real state of the economy looked less and less like his theoretical one. The structure Smith erected was an ideal, the economic counterpart of geometry's ideal figures, Descartes' hypothetical ideal human body, or an ideal frictionless machine. Unfortunately, none of these things exist in nature. Smith made the mistake of supposing that the perfectly balanced, ideal economy is an exception, that it really exists—and therein lies the flaw of his theory. The economy cannot approach this ideal without some legislative oil to keep the wheels turning.

To men who were still influenced by their Christian heritage, it seemed that human reason and the mathematical method could discover the same moral and ethical laws that Christianity revealed. Thus, even the moral sphere could be made "scientific" and a Heaven on Earth erected by rational mortals. The flood tide of science became the ebb tide of religion and "the sea of faith" retreated with "a melancholy, long, withdrawing roar . . . down the vast edges drear and naked shingles of the world."

Truly, the mathematical tree with its method of rigorous, logical thinking seemed to provide the answer to all men's problems, in the social as well as the physical and natural sciences.

In the year that George Washington began his second term as President, a baby boy was born in Russia, an infant who would grow up to chop down the mathematical tree, just as Washington, according to legend, chopped down the cherry tree. And not only he, but all mathematicians, would be left saying, "I cannot tell a lie, *but what is the truth?*" To labor the metaphor of the mathematical tree a bit further, for over two thousand years the world had accepted Euclid so completely that they could not see the forest for the trees. The obvious had been staring them in the face for centuries, but no one saw it. Nicholas Lobatchevsky was the first—or one of the first.

Lobatchevsky was born in northwest Russia on November 2,

1793, of Polish parentage. His father, either an architect or a surveyor—no one knows for sure—had migrated to Russia several years earlier.

When Nicholas was only three, his father died and Praskowja Ivanovna took her three small children, Alexander, Nicholas and Alexei, south to Kazan. The family was extremely poor, but Praskowja, a cultured, educated woman, was determined that her children would not be stifled by poverty and ignorance. She herself became their teacher until they were old enough to go to the local elementary schools. All three boys were accepted as scholarship students.

Nicholas was eight when he entered school and quickly impressed his teachers with his brilliance, especially in Latin and mathematics. He raced through the elementary courses and at thirteen won a scholarship to the newly founded Kazan University, where his older brother Alexander was already enrolled.

During Nicholas' first two years, the university faculty was understaffed and generally undertrained. Slowly more and more professors were brought in from Germany, where teaching standards were among the highest in the world. Since only one of the German professors at Kazan could speak Russian and few of the students could speak German, there was a difficulty that today would be insurmountable, but which was solved then by having the teachers lecture in Latin—at that time still the international language of all educated people.

Russia continued to follow Peter the Great's program, trying to drag itself up to the level of western Europe. But it could not afford to import teachers forever. It had to develop its own, and as part of the program, granted scholarships on the condition that the recipients become teachers. Nicholas and his two brothers won scholarships to the university on this basis, pledging themselves to teach for six years following graduation. Nicholas and Alexei both fulfilled the promise. Alexander, for unknown reasons, committed suicide before he was graduated.

Although Nicholas enrolled as a medical student, it is certain that he continued to study mathematics, for in 1809 he was made *Kammerstudenten,* or honor student, in mathematics. He spent several hours a week of independent study under that maker of master mathematicians, Johann Bartels, who had been one of the first teachers of Carl Gauss.

Until Nicholas was made *Kammerstudenten,* he had been an ideal student, but the honor suddenly raised his confidence and lowered his inhibitions. With high-spirited self-assurance, he engineered all kinds of pranks, made trouble in class by teasing his teachers, and scared the wits out of everybody by setting off firecrackers late at night. One teacher in particular complained about him so much that Nicholas eventually lost his honorary position. That he was not more severely punished or even expelled was due only to the intervention of Bartels, who knew brilliance when he saw it.

At the age when most boys are only beginning college, Nicholas was eligible for his Master's degree. The university, however, had had enough of his practical jokes and bad behavior and decided to withhold the degree. Nicholas apologized and promised to be good; the university relented—after all, Lobatchevsky was their most brilliant student and it might be better having him as an assistant professor than as a classroom clown for another year.

And so, Nicholas at eighteen began his long career of teaching. He had pledged six years to the university. He served forty. Kazan had given him ten years' schooling during his poverty-stricken childhood. In return, he gave everything of himself not only to the university but to all the schools in the city. He served diligently and well. By the time he was twenty-one he was in charge of the money spent in all the Kazan schools—elementary, intermediate, and university. At the age of twenty-eight, he was a full professor teaching all the classes in mathematics and astronomy as well as theoretical physics. In addition, he was still treasurer for the schools and university librarian, practically running the library-museum singlehanded. His predecessor had left this in such a terrible state that it was almost impossible to find anything. Some years before, his brother, Alexei, had sent the university a collection of rocks from Siberia, but the whole collection had been lost before it was ever exhibited. In sorting out the crates of books, Nicholas came across his brother's rock collection —not even unpacked. There was enough work in clearing up the confusion in the library to keep several men busy—but Nicholas did it alone. He sorted, catalogued, and shelved hundreds of volumes. There is a story that one day while he was sorting books, a visitor came to the library to look around. Nicholas guided him through the various rooms, pointing out the collections and commenting on new editions. After the tour the visitor thanked him, and thinking him to be the janitor, offered him a tip for his services.

During his years on the staff, Lobatchevsky managed to remain aloof from the petty quarrels and politicking that went on among the rest of the faculty—neutrality was no easy trick in a university whose rector did his best to make the worst of every situation. Almost all the German professors left, disgusted with the domineering, overbearing ways of the administration. Religious differences—the Germans inclined toward atheism—had only heightened the other differences.

The exodus of so many teachers meant that those who remained had to work twice as hard—which accounts for Lobatchevsky's almost impossibly heavy schedule. Some professors were even teaching two classes at once. It also meant a chance of rapid promotion for anyone with ability. As a result, at thirty-three, Lobatchevsky, the janitor-professor, was made rector or head of the university. He took this position as seriously as he had taken the others. When new buildings were to be added, he studied architecture so that he could supervise the project knowledgeably and, for perhaps the first time in history, a state building project was completed with less than the allotted money.

Under Lobatchevsky the university prospered and grew. The faculty was enlarged and standards raised. Then, in 1830, all the advances were threatened with destruction. Cholera came to Kazan—deadly, horrible, it swept over Europe, killing hundreds of thousands. Sometimes entire towns were almost depopulated. Yet it was Lobatchevsky's duty to try to save his students and staff. He went to see the Governor of Kazan to discuss measures of keeping the epidemic under control but found the city gates locked—the Governor's first precautionary measure. Lobatchevsky hurried back to the University, ordered the faculty to bring their families to the campus, and then had the university gates locked. No one except doctors was permitted to enter; no books were allowed in, for it was thought that the disease might be carried in the bindings. Loose manuscript pages, however, were allowed to pass through the gates after being disinfected with steam. Lobatchevsky's rigid rules about sanitation—at a time when men did not even know of the existence of germs—lasted until the plague had spent itself in six weeks. The University had forty cases of cholera with only sixteen deaths, a mortality rate of about 3%.

Besides running the University and several of its departments, and editing the University journal, Lobatchevsky was a one-man talent scout in ferreting out deserving students. In his own quiet, earnest way, he took a fatherly interest in his students and was always ready to help a bright boy through school. One day while shopping he noticed that the young clerk who waited on him was reading a book on mathematics. Suspecting that the boy was wasting his time working in a store, Lobatchevsky helped him to be admitted to the University where the ex-clerk later became a professor of physics.

In 1832, Lobatchevsky managed to find time from his professional duties to attend to private ones. For more than half his life he had literally lived for the University, devoting all his time and energies to it. The most pressing problems that attend any young institution had now been solved. He could relax a bit, work on his mathematical projects, enjoy a fuller social life, and begin a family. He married at the age of forty and within the next few years had four children, three boys and a girl.

His work as rector, while not routine, was at least organized and no longer required all his time. Yet fate seemed determined that Lobatchevsky should not rest. In 1842 a fire broke out in Kazan. When it was finally brought under control, half the city lay in ashes, including several University buildings, among them the observatory which had been under Lobatchevsky's personal care. Again Lobatchevsky threw himself into his work and by the end of two years the astronomy buildings had been completely rebuilt and furnished with the equipment that he himself had rescued from the flames.

His superb administrative abilities were brought to the Tsar's attention. In 1833 Tsar Nicholas awarded him a diamond ring for his excellent work; later he bestowed several medals and honorary awards on him for other outstanding services. Apparently Lobatchevsky's hardheaded executive ability and practicality carried over even into his personal life, for although a ring from the Tsar had an honorary and sentimental value, to Lobatchevsky the financial value was enough to induce him to sell the jewel in order to buy some prize sheep for his farm.

Admirable though his successes as an educator and administrator were, they are negligible compared to his achievements in mathematics. The work for which he was honored and famed in his lifetime is now forgotten. Who knows or cares that it was Lobatchevsky who introduced physical education into Russian schools, developed an obscure university into a respected seat of learning, or instituted a program of adult education for Kazan's laborers? For all these achievements which seemed so grand a hundred years ago, Lobatchevsky now rates a mere footnote in the history of school officialdom. It is for other work—work which at the time was ignored—that he wins a whole chapter in the history of mathematics and a seat of honor among the great.

His work begins with Euclid and the not so "self-evident" axiom that through a point in a given plane only one line can be drawn parallel to another line on the plane. Euclid himself may have been disturbed about this axiom, for he delayed using it as long as possible. One modern mathematician has gone so far as to state that Euclid's greatness lies mainly in his suspicion of the axiom. It may be, however, that he was not suspicious at all but deferred using it for the sake of elegance. At any rate, for hundreds of years this axiom bothered other men and by the eighteenth century was "the scandal of geometry." It was not the truth of the theorem that was doubted, for in two thousand years its inclusion had never led to a single contradiction or inconsistency, which would almost surely have happened if the axiom were false. It was the "self-evidence" of the axiom that was disputed. Yet all attempts to deduce the axiom from the others—that is, to prove it as a theorem—or to find a more agreeable substitute failed.

Not until the early eighteenth century was any progress made. An Italian Jesuit by the name of Saccheri hit on the right solution but was so dismayed at what he found that he concluded that Euclid's axiom must be the only possible one. Completely ignoring his own investigations, Saccheri published his juggled results in a book inappropriately entitled *Euclid Vindicated from All Defects*.

Lobatchevsky, too, was intrigued by the two-thousand-year failure to put the parallel axiom on a sounder basis. Like all his predecessors, he first tried to prove the axiom as a theorem, although he did not believe it could be done: "The fruitlessness of the attempts made since Euclid's time . . . aroused in me the suspicion that the truth which it was desired to prove was not contained in the data themselves." In 1825 he published his unstartling finding: the axiom cannot be proved.

The problem still rankled. If the axiom could not be proved, why not try to disprove it? Or why not try to find a substitute? The ideas melded into one and Lobatchevsky proceeded along the same lines as Saccheri. If the axiom were assumed to be false, then either no lines through a point parallel to another line, or more than one line, could be drawn.

He tried constructing geometries using either of these assumptions plus Euclid's other nine axioms. When the assumption that

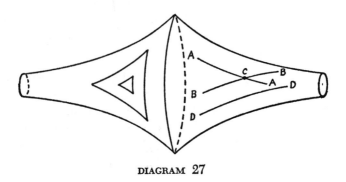

no lines could be drawn through the point was used, inconsistencies resulted. Obviously the assumption was false. But when he assumed that more than one line could be drawn through a point, he found no contradictions. The theorems derived from this set of axioms were indeed strange—strange enough to have frightened off Saccheri, but not Lobatchevsky. In this new geometry, for example, the sum of the angles in a triangle is less than 180 degrees; the smaller the triangle in area, the closer to 180 degrees is the sum of its angles; and only congruent triangles are similar.

This non-Euclidean geometry—"pangeometry" or "Imaginary Geometry," the names Lobatchevsky used—cannot be depicted on the ordinary flat plane of Euclid. Instead, it can be visualized on a pseudosphere (Diagram 27). On the pseudosphere, more than one line (A and B) can be drawn through the point C parallel to line D. (Actually an infinite set of lines can be drawn, but two will suffice as an illustration.) No matter how far the lines A and B are extended, they will never meet line D. The diagram also contains two triangles—one large and one small. The sum of the angles of the smaller triangle is greater than that of the larger triangle, yet both are equilateral triangles and therefore, according to Euclidean geometry, should be similar.

This imaginary geometry is as logical and precise as Euclid's and has as much right to the title of Truth as does Euclid's. Lobatchevsky's "straight lines" may seem to be curves and his "plane" not a plane at all. Yet he uses the same definitions for these concepts as does Euclid. It is simply tradition that has ossified men's thinking in this matter, for there is nothing in Euclid to indicate that "a surface which has the property that every straight line which joins any two points in it lies altogether in the surface," is the conventional plane and not Lobatchevsky's pseudosphere. This definition, supposedly precise and absolute, seems so only when one knows in advance what a plane is. Actually, the definition is open to a number of interpretations. For centuries men visualized the conventional plane when confronted with Euclid's definition and based their whole geometry on this unwarranted picture. But geometry is not based on pictures. It is based on logic. Diagrams are nothing but schematic renderings of abstract relationships and have no bearing on the validity of the relationships themselves. Thus, from the point of logic, there is no more reason to accept Euclid's geometry than Lobatchevsky's.

From the point of usefulness, however, Euclid's geometry appears more acceptable. For two thousand years men had used it in studying the Earth and the universe, and it had proven itself over and over. Certainly nothing could be a more accurate model of physical space. Euclidean geometry was capable of describing everything from paths of heavenly bodies to the sizes and shapes of specks of dust. If space is Euclidean, then it cannot also be non-Euclidean and the new geometry is a useless model of a nonexistent kind of space. It has been erroneously asserted that Lobatchevsky himself did not think of his new "imaginary" geometry as anything but a mathematical oddity which, though logical, was not descriptive of our universe. Actually, Lobatchevsky believed that his new geometry might furnish a better description than Euclid's. He took the highly novel view that geometry is an empirical science rather than a purely theoretical one, stating as early as 1825 that to prove or disprove the parallel axiom "the aid of experiment would be needed, for example, astronomical observations, as is the case with other laws of nature." Such a proposal must have struck his contemporaries as about as absurd as testing the truth of $2 + 2 = 4$ by observing what happens when two of anything are added to two others of the same thing. Gauss, however, did conduct a test using terrestrial triangles to try to determine which was more accurate, Euclidean or non-Euclidean geometry. The results were inconclusive.

The sudden appearance of a new kind of geometry did not cause men to throw away beliefs supported by thousands of years of tradition even though these beliefs did not always correspond to the evidence of one's very eyes. Euclidean geometry assumes that lines can be made infinite—yet who has ever seen this? It assumes that parallel lines never meet, yet a glance down a railroad track indicates just the opposite. It assumes that space and time are separate entities, yet the paradoxes stemming from this assumption were noted by Zeno as early as the fifth century B.C.

The flaws in a Euclidean interpretation of outer space were not obvious to early astronomers, but by the nineteenth century unaccountable discrepancies between what was and what should be plagued the whole Newtonian system. Most notable was that the path of Mercury did not jibe with mathematical calculations. Furthermore, the explanation of the behavior of all bodies depended on a fictional, or at best inexplicable, force called gravity. All these problems made Newtonian physics, which is based on an Euclidean interpretation of space, appear suspiciously over-simplified.

A famous mathematician, Lagrange, had once remarked how fortunate Newton was, for only one man could discover the system of the universe—and next to this achievement, all others paled. The sentiment is correct; the statement is false. Newton's "System of the World" was based on Euclidean geometry. He looked at space and time as absolute and infinite. Space to him was a three-dimensional entity; time, a separate one-dimensional entity. This way of looking at the world seemed to be the only possible way. Men were so certain of these assumptions that they never even thought of them as assumptions. They were truths, uncontestable, absolute, and self-evident.

More than half a century passed between the discovery of non-Euclidean geometry and the brilliant idea of applying this geometry to outer space. At the beginning of the twentieth century there appeared another genius who, like Newton, would erect a system of the universe—a new system based on non-Euclidean geometry. His name was Albert Einstein.

Einstein's system differs from Newton's in three major respects: Space is non-Euclidean, space is not absolute, and space is not a separate entity from time. The geometry of space is not Euclidean, according to Einstein, nor is it even a uniform non-Euclidean system. The presence of matter alters the properties of space in such a way that no one geometry can describe it. That is, the geometry on Earth is not the same as that, for instance, on the Sun. Space varies in its geometrical properties, just as Pascal's projected figures varied from the original or Lobatchevsky's larger isosceles triangles differ from his smaller ones. If a man on Earth were to measure a certain distance or period of time on Mars, he would obtain different measurements from those of a man on Mars measuring the same distance or time. The discrepancies would occur not from errors or optical illusions but because each man's measurements are based on his own local length and local time. Yet just as certain properties remain invariant in projective and non-Euclidean geometry, so certain properties remain invariant in the different space-time worlds. Euclid's—and Newton's—three dimensions have been replaced in Einstein's system by four—the fourth being time. Thus Einstein maintained that viewing time and space as different and separate entities is an artificial distinction imposed by man.

One of the most important aspects of this new view of the universe is that it dispenses with the fiction of gravity that Newton introduced to explain the course of planets, stars, and falling objects. The new theory asserts that all objects move in straight lines—non-Euclidean straight lines. Thus when planets move in elliptical orbits around the Sun they are simply moving in the "straight lines" of their particular geometry.

Furthermore, Einstein's more complex theory is, in application, simpler than Newton's. That is, it describes the way the universe works without having to allow or explain away a mass of detailed exceptions. While Newton's theory is easier to understand in that it can be easily visualized, the theory becomes extremely complicated in practice. Also, Newton's explanation that the universe operates like a machine simply begs the question, for how does a machine operate? One is left, like Byron, wondering "who will explain the explanation."

No complete discussion of Einstein's theory of relativity is intended here. What is intended is to illustrate non-Euclidean geometry's ability to describe space as well as, if not better than, Euclidean geometry's, and to hint at some of its far-reaching applications and of the truly revolutionary theory to which it has given rise.

Lobatchevsky's view of geometry as an empirical science has been vindicated in the Einstein theory. By observing, collecting data, and *then* selecting the geometry that best fits the data Einstein formulated his theory. Several different geometries or "space models" were at his disposal, but he did not assume a priori that any one of these was the correct one. Actually the one he chose was not Lobatchevsky's but another invented later by Riemann. It is interesting to note that Lobatchevsky's suggestion that the truth of Euclidean or non-Euclidean geometry might be proved by astronomical observation was correct.

A parallel between geometry and the inductive sciences can now be drawn. Science, once thought to be absolute truth, was forced in the eighteenth century to retreat to a position where it claimed to be only a *technique* for finding highly probable truth.* With the rise of non-Euclidean geometry, geometry also was forced to retreat from its lofty citadel of absolute truth to a lower level where it took its place as only one step in a process of discovering probable truth. And as geometry went, so went the rest of mathematics. It is not God who geometrizes, as Plato said, or arithmetizes, as Jacobi said, but man. Mathematics is no longer an absolute truth; there is no objective reality, no *pi* in the sky. Mathematics is only a very useful tool—and like any other tool is man-made. The world does not necessarily fit our mathematics. On the contrary, we fit our mathematics to the world, observing and testing to determine which mathematics to use under different circumstances in order to have the world as we see it conform to some type of mathematical order. Mathematicians do not suppose that the mathematics chosen is a perfect fit any more than a suit selected in a store fits a customer perfectly. Nor do they make the old mistake of supposing that the mathematical model resembles or mirrors the real world any more than a pair of trousers resembles a man. It was the Greeks who first conceived of the world as being made in a mathematical mold—a conception that is still held by most people. Mathematicians now, after two millennia have come to realize that such a view is as erroneous as imagining that men were made with two legs so that they would be able to wear trousers.

The main question now is: Is mathematics man-made to describe and fit the world—that is, are its axioms based ultimately

* See chapter on Leonhard Euler.

on something either empirical or intuitive—or is mathematics man-
made and then the mathematical model chosen that best fits the
world—that is, is mathematics a purely mental creation, a consistent, self-contained system drawing on no material from outside itself? Putting this question into an analogy, are trousers
made after looking at a customer to determine what kind of
clothing he can wear; or are trousers selected from an already
existing stock of clothing and then given to the customer with
two legs?

This question has divided modern mathematicians into two
camps, the intuitionists and the formalists. The intuitionist is
not, as one might suspect from the name, a person who depends
on intuition in solving problems. Quite the contrary, he is extremely rigorous and rejects much that a formalist would accept.
He is the man who insists on tailored trousers. The intuitionist
maintains that it is impossible to erect a rigorous, complete,
logical, and self-contained mathematical system, that eventually
we reach the point where we must appeal to intuition as a basis
for the axioms. The formalist, on the other hand, maintains that
mathematics can be made into a formal, logical system that depends only on logic and a consistent set of axioms. A formalist, as
might be suspected, is more liberal than an intuitionist in his interpretation of what constitutes a valid proof. He will accept the
ready-to-wear trousers. These arguments between the two schools
of thought will be resumed in the final chapter on Cantor.

Other philosophical arguments were revived by non-Euclidean
geometry, especially in connection with the Einstein theory. Scientists—indeed all humans—assume that cause and effect exist. It
is part of our whole ordering process. Furthermore, we assume
that cause always precedes effect. If a man puts his hand in a fire,
he will burn himself. But he cannot burn himself before putting
his hand in the fire. The cause—putting his hand in the fire—
must occur before the effect—burning. Yet Einstein's theory of
relativity asserts that two such events—putting one's hand in a

fire and being burned—may appear in different order to observers living in different space-time worlds. One observer may see the man's hand being burned *before* it is put in the fire. Thus, in rejecting absolute space and time, absolute causality is put in jeopardy, for it *seems* (and *is* to us poor mortals) impossible for effect to precede cause, yet this is precisely what can happen in relative space-time worlds. And if there is no cause and effect, there can be no determinism. A similar breakdown of causality occurs at the other end of the scale—in the microcosmic world of quantum physics. It is only in between, in the "seeable-sized" world, that causality appears to operate absolutely.

The problem of determinism is not settled—and perhaps never will be. Many eminent physicists—especially quantum physicists —now reject a rigorous causality, yet Einstein believed that absolute causality does exist and that the current paradoxes will be resolved as science learns more. In many ways these determinists, with their faith in a rational system, correspond to the formalists of mathematics while the nondeterminists, who hold that a rigorous causal description of nature is impossible, correspond to the intuitionists.

This faith in a rational universe in which cause and effect operate in many ways parallels the medieval faith in God, which it replaced. Today, modern man finds comfort in the idea of an ordered universe—nor does his comfort necessarily depend on the belief that this order was put there by a higher being. A rational universe offers us security, intellectual challenge, and the power to control our environment. An irrational universe seems to offer nothing but chaos. No wonder men cling to their faith in causality. Yet science, when it first challenged religion, seemed to offer little comfort to man. It took away his Heaven, upset the universe of which he was the center, and gave him ancestors with tails. He managed to survive these indignities and erect a satisfying world. Perhaps the offerings of quantum physics, the theory of relativity, and non-Euclidean geometry may prove to be equally

as satisfying. Indeed, this veritable revolution in human thought
that was initiated by Lobatchesky's discovery has already given
man something to be proud of. Gauss had speculated that there
might be superior beings who would look down on those who
could conceive of only three dimensions. These superior beings
now exist—and they are men, men viewing a four-dimensional
space-time world.

"What Vesalius was to Galen, what Copernicus was to Ptolemy,
that was Lobatchevsky to Euclid," stated one English mathematician. Yet Lobatchevsky received neither the respect that had
been accorded Vesalius nor the ridicule heaped on Copernicus.
His startling geometry was first published in 1826. Its reception
was just as startling: No one paid any attention. Sixteen years
later Gauss learned of his work and prodded the Royal Society
of Göttingen to admit him as a corresponding member. But still
no recognition came. In Kazan—indeed, in all of Russia—Lobatchevsky never won a single mention for his greatest achievement.
It was not until ten years after his death that mathematicians began to notice non-Euclidean geometry. Ironically, their attention
was focused on the subject not by a rediscovery of Lobatchesky's
work but by the revelation in Gauss's diaries that he himself had
investigated non-Euclidean geometry.

Even Lobatchevsky's less ambitious mathematical projects fared
badly. An ordinary geometry textbook he had written was rejected by a so-called expert who remarked, "If the author thinks
his work can be used as a textbook, it is obvious that he has no
idea of what a text should be . . . nor any comprehension of
geometry." Lobatchevsky apparently did not take his critics seriously but regretted wasting so much time on something that came
to nought.

Because Lobatchevsky had begun his career at such an early
age and married so late, his sons and daughter were still children by the time he was forced to retire in 1846. Although he was
only fifty-three when he stepped down from the rector's chair at

the University, still able and willing to devote himself to his work, government policy decreed that he had served the maximum number of years. His colleagues protested that he was too young to retire and furthermore, that no one could possibly replace him. But the wheels of bureaucratic red tape ground on and put an end to Lobatchevsky's career. He had won medals and rings, recognition and respect, but was denied the real reward: being allowed to continue in his work.

The blow of forced retirement at the prime of his life was staggering. It was followed by another that crippled. Lobatchevsky's oldest son died suddenly. The poor man could not even find relief from his grief in work as Gauss had. His hair began to turn white, his shoulders sagged like those of an old man, and his sight began to fade. He was still allowed to teach but very seldom went to the University except at examination time to hear the dissertations of graduate students.

Although his major work—non-Euclidean geometry—had been completely neglected by others, Lobatchevsky himself realized its importance and continued to work on it, in spite of increasing blindness. Like Euler, he dictated everything to a secretary. In 1855 his final paper on non-Euclidean geometry was published as part of the celebration of the University's fiftieth anniversary.

Lobatchevsky's main pleasure during his last years came from working on his farm about fifty miles from Kazan. There he raised prize sheep and discovered a new way of processing wool, for which he won a government citation. At the age of sixty-one he set out a grove of trees and was heard to remark that he would not live to see them bear fruit. On February 24, 1856, only a few months after his prediction, he died and was buried in the newly planted orchard.

Évariste Galois
1811-1832

Madison was President of the United States; an insane king, George III, sat on the English throne; and in France Napoleon strutted before his bedazzled court. The year was 1811, and all was not right with the world. Bloodshed and slaughter would soon settle the issues. It was in this year that Évariste Galois was born, and in his short life, he personally reflected all the ills of the age.

At one o'clock on the twenty-sixth of October, 1811, Nicholas Gabriel Galois, director of a boarding school in Bourg-la-Reine, brought a baby boy, one day old, to the mayor's office to register the birth of his first son. Adelaide-Marie Demante Galois and her husband named their child Évariste.

The Galois family was a happy and respected one. Madame Galois was well-educated, clever, and intellectually sophisticated. Her husband was a charming, fun-loving man who delighted in making up clever verses about his friends. The citizens of Bourg-la-Reine expressed their admiration for him by electing him mayor shortly after Évariste's birth.

Like most middle-class children, Évariste, his older sister, and a younger brother, Alfred, were tutored at home by their mother until they were ready for school. At the age of twelve, Évariste passed his entrance examinations and was admitted to Louis-le-

Grand, a boarding school in Paris that had been attended by Voltaire and Victor Hugo.

Despite its illustrious alumni, Louis-le-Grand seems to have been dedicated to mediocrity and misery. The students were wakened at five-thirty every morning to dress in dark, unheated dormitories that were crowded with bug-infested beds. Discipline was undiluted with any milk of human kindness, and the food was hardly the type for which France is famous. Classes began early in the morning and continued until evening, with an undue emphasis on Latin and the classics. If the boys could read and write Latin fluently, they were considered educated.

As can be imagined, the students did not easily adapt to this grueling regime. To be more specific, their hatred of the teachers, the school, and its director was so intense that during Évariste's first year the students rebelled. At a given hour they attacked, throwing dictionaries, textbooks, and Latin and Greek grammars at the heads of their instructors. The insurrection was put down, forty student leaders expelled, and the old harshness and grinding continued.

The day after the fracas, a dinner was held at which a hundred and fifty of the school's best students were honored. Here again the pupils showered their wrath and hatred on the men who ran the school. The boys refused to join in a toast to King Louis XVIII, who had ascended the throne after Napoleon's exile, and sat sullen and silent during the toast to their school director. After the meal was over, they proceeded to create a near riot by singing the outlawed Marseillaise, anthem of the Republic that had been dead for a number of years.

The school authorities dealt with this insolence, insubordination, and virtual political treason by expelling its hundred and fifty brightest students. Unfortunately, Évariste was not lucky enough to be among them, for he was one of the slowest boys in the school—so slow, in fact, that when he was fifteen he was demoted.

Forced to repeat another year filled with dreary Latin declensions and conjugations, Évariste decided to vary his studies by electing mathematics—a subject that was not part of the compulsory curriculum—and joined the geometry class. It was his first exposure to mathematics and it thrilled him. Suddenly be became transformed, as though he were in the grip of a demon. Nothing mattered anymore, nothing but mathematics. Every minute of his day was devoted to studying and thinking mathematics. He raced through the geometry book in two days, enthralled by the beautiful, rational structure of the axioms and theorems. Ordinarily, students completed the book in a two-year course of study. Évariste ransacked the library for other volumes on mathematics. He found an algebra book and began teaching himself from it, but soon threw it down in disgust. The book was a commonplace text that presented the subject as a hodgepodge of rules and procedures. Évariste turned to the masters for instruction—Legendre, whose geometry he had read, Cauchy and Lagrange. He read works meant for mature mathematicians, for men who had been working in the field for years—not days or weeks. And yet, Galois understood and absorbed it all.

Disturbed that algebra was neither as elegant nor as complete as geometry, he decided to remedy matters. One problem that plagued him was the solution of equations above the fourth degree. Equations of lower degrees could be solved by one means or another. When the usual methods failed, special formulas could be used, such as the quadratic formula for equations of the second degree:

$$x = \frac{-b \pm \sqrt{b^2 - 4ac}}{2a}.$$

But no such formula had been devised for the solution of equations above the fourth degree. Galois set himself the task of finding it. Still a schoolboy with not even a whole year of mathematical training, he was tackling a problem that had stumped the greatest mathematicians since Cardano. Finally Évariste correctly decided—although he had no proof—that fifth-degree equations are insoluble by radicals, i.e., by a formula such as the quadratic one, and turned his attention to the more basic problem of determining whether such equations are soluble at all.

By the end of the school year, Évariste's genius for mathematics would have stunned an ordinary teacher. But the teachers at Louis-le-Grand were not ordinary. They were either too arrogant or ignorant to notice the genius fate had put in their hands. One teacher wrote in his report on Évariste: "A character whose traits I do not understand, but I can see that it is dominated by conceit. . . . He has neglected most of his work in class. This is the reason why he did not do well in the examinations."

Another said: "His ability in which one is expected to believe but of which I have not yet witnessed a single proof will lead him nowhere. There is no trace in his work of anything but queerness and negligence."

Another, in whose eyes Évariste was anything but a model student: "Always busy with things he ought not to do. Gets worse every day."

So much for the insight of his teachers.

The "things he ought not to do" were mathematical problems, which he worked at incessantly. Mathematics was the only thing that mattered to him. One of his teachers, possessed of more sense than the others, suggested that it might be better if Évariste dropped his other subjects and studied mathematics exclusively, for "a mathematical madness dominates the boy. He is wasting his time here and only torments his teachers and constantly is punished. His conduct is very bad, his character secretive."

Évariste's mathematics teacher—a pedant if ever there was one —was distressed that the young boy did not plod along with the rest of the class: "Intelligence and progress marked; but not enough method," he commented. By "method," he meant that Évariste did not write out each consecutive step of a solution. The young genius could make lightning calculations in his head and found it annoying to write down each obvious detail.

Évariste counted the days until he would be free of Louis-le-Grand and could go to the École Polytechnique, the best school of mathematics in all of France. There he would surely be appreciated. His teacher at Louis-le-Grand, however, felt that he should stay on for another year to improve his "method." Évariste saw no point in it—he already knew more than his teacher—so, at the age of sixteen, he took the Polytechnique admissions examination. The questions were easy. He knew all the answers. But Évariste had a terrible enemy whom he was powerless to fight and who followed him all his life. And now that enemy caused him to fail the examination. The enemy was himself—an enemy he could neither conquer nor escape. Évariste was insolent to his examiner and openly contemptuous of the questions. The result was that he spent another year at Louis-le-Grand under the I-told-you-so eyes of his teacher.

During this time he wrote a paper on his researches into the solvability of algebraic equations, which he submitted to the French Academy. According to one writer, this paper contained "some of the greatest mathematical ideas of the century," and the nineteenth century was an extremely prolific and fertile one for mathematics. Évariste had written his paper when he was only seventeen. He sat back and waited.

And while he waited, a tragedy occurred a few blocks from the school. The mayor of Bourg-la-Reine, Nicholas Galois, left home, made the short trip to Paris, and killed himself. For a decade and a half he had weathered all the political ups and downs: Napoleon's Empire, a provisional government, the Hundred Days after Napoleon's return from Elba, and the restoration of the Bourbons to the throne of France. All this was not easy in a country torn by conflicting political factions and religious strife, with the Jesuits the archenemies of the Republicans and Protestants. Évariste's father had accused the Jesuits of trying to undermine him. Someone had circulated obscene verses among the towns-people of Bourg-la-Reine, verses purported to by Mayor Galois, whose penchant for rhyme was well known. Slanderous rumors had been circulated until the poor man's reputation decreased to the point where people considered him to be the town crackpot. In despair he committed suicide.

Évariste went home for the funeral—a desperate, grief-stricken boy, filled with hate for the parish priest who had persecuted his father. At the grave Évariste accused the Jesuit of having murdered his father, and the mourners showed their agreement by hurling stones and sticks at the retreating priest. Évariste threw himself on his father's coffin and sobbed. Another link in the chain of hate had been forged. His teachers, the priest—he hated them all.

His bitterness over the death of his father was soon directed toward the French Academy where he had sent his paper. Évariste waited for months and when he still heard nothing, inquired and was told that the paper had been sent to M. Cauchy, France's most eminent mathematician. Further inquiries revealed that M. Cauchy knew nothing of the paper. He denied ever having received it. Évariste speculated—probably correctly—that Cauchy had thrown the paper into the wastebasket. If so, Cauchy's wastebasket was the receptacle of some of the greatest mathematical works of the nineteenth century, for this was not the first time he had mislaid the work of an unknown. A young Norwegian, Niels Abel, had also sent him a paper on algebraic equations of the fifth degree, which like Galois' had disappeared.

During his sixth year of the four-year course at Louis-le-Grand Évariste took the entrance examination to the École Polytechnique and again failed. Impatient with the idiocy of his examiner, who, like his teachers, stressed method more than imagination, Évariste destroyed all his chances by flinging an eraser at the examiner and hitting him in the head. "This would be my answer

to your question, sir," he said, crowning his deed with imperti-
nence.

While a blow on the head, according to legend, was beneficial
for Newton, it was just the opposite for Galois. The doors to the
École Polytechnique were now closed to him forever. He finished
out the term and left Louis-le-Grand at last. He had impressed
his instructors with his strangeness, his lack of sociability, and his
rebelliousness, but not with his genius. The final remarks about
him were: "His character is strange and he pretends to be more
strange than he really is." His mathematics teacher dismissed him
curtly with: "Conduct good, work satisfactory."

With luck—for no one seems to have recognized his intellectual
abilities—he was admitted to the École Normal, an institution of
lesser quality than the École Polytechnique. The caliber of its
teachers can be guessed from the following report on Évariste:
"He is the only student who answered me badly; he knows absolutely nothing. I have been told that he has mathematical ability;
this certainly astonishes me. Judging by examination, he seems of
little intelligence or has hidden it so well that I found it impossible
to detect."

Évariste's teachers not only did not appreciate him, they ac-
tively disliked him. His conceit, coupled with a strange fanaticism
and air of persecution, was hardly an appealing quality. One day
in class the teacher stated a new theorem in algebra that had
been just proved by a European mathematician, although the
proof had not yet been published. Évariste sat in his seat with
his usual contemptuous expression on his face. In an obvious attempt to humiliate, the teacher invited him to the blackboard to
prove—if he could—the theorem. Évariste wrote out the proof to
the astonishment and chagrin of the teacher who, by now, one
would expect, would have recognized his pupil's genius. That
he did not may be partly Évariste's fault, for he had the remarkable knack of reducing his intellectual successes to personal vic-
tories, with the result that his brilliance was viewed merely as

conceit. Every intellectual achievement made him more and more disliked by his teachers and Évariste, in turn, frustrated by lack of recognition, became more withdrawn and smug. Only one person, a fellow student named Auguste Chevalier, seemed to realize that Galois' genius was greater than his faults.

Meanwhile, this strange eighteen-year-old began writing his theories on permutations and combinations and algebraic equations and submitting them to various mathematical publications. Several very short pieces were published but went unnoticed. A more complete account of his theories was submitted in the annual contest held by the French Academy of Sciences. Six months later, when he had received no word about this paper, Évariste made inquiries. Once more he learned that his paper had been lost. That his work, which today forms the basis of modern algebra, should be completely ignored, carelessly misplaced, seemed incomprehensible to the young mathematician.

The stupidity, narrow-mindedness, ignorance, jealousy, and indifference of others bred in him equally negative emotions. He began to hate the Academy, the men in it, and all it stood for. His hatred spilled over and spread to the political area. The existing order, the state of society, rather than individuals, were to blame for his troubles. Galois became an avid Republican. Like most fanatics, he was driven more by hate than by love. The monarchy became the focus of all his bitterness against his overbearing teachers, against the priest at Bourg-la-Reine who had driven his father to suicide, against the wise men of the Academy who lost his papers, and against his classmates who ridiculed him. Under a Republic all this would be changed.

Ever since Napoleon's exile, revolution had been in the air. The people wanted a republic, not a king. Then, in 1830, the revolution came and the long-awaited Republic was ready to be born. France waited hopefully—and in vain.

During the revolt, Galois was at the École Normal, where he tried to incite his fellow students to join in the uprising. The director, hearing of his actions, had him locked in his room until the fighting was over. Yet when the revolution was a *fait accompli*, the director wholeheartedly embraced the new provisional government. Some months later one of the students at the school wrote an anonymous letter to the newspapers exposing this hypocrisy. There was no doubt as to who had written the letter, at least not in M. Guigniault's, the director's, mind. Who else but Galois? He was expelled immediately, and there his formal education ended. Évariste was nineteen years old.

At last he was free. But France was not. The revolution had removed one king and in short order another had been installed, the Duc d'Orléans, who now reigned as King Louis Philippe. Galois joined the artillery of the National Guard, made up of other Republicans like himself, and worked and dreamed of the future. The Republicans bided their time until the new king and his none too stable government could be overthrown and a true democracy instituted. The moment seemed to have arrived on December 21, 1830. The National Guard stationed itself in the Louvre—today a museum, at one time a royal palace. They planned to used the building as a fortress in the ensuing revolt. But the French people were tired of fighting and bloodshed. The king promised them liberty—and for the sake of peace, they chose to believe him. Only the most dedicated Republicans wanted to fight, and the revolt fizzled out. The Artillery Guardsmen—Galois among them—laid down their arms and went home when they discovered that the people were not behind them.

A few weeks later, in January, 1831, Galois began giving a private course in algebra, dealing with imaginary numbers, the theory of equations, the theory of numbers, and elliptic functions. Quite a number of students turned out for this first lecture—about forty in all. The following week only about ten students came and by the third week the number had dwindled to four. After that even Galois did not appear. So ended his short-lived teaching career.

He was now drifting about in Paris, still studying mathematics by himself, working with the Republicans, and living a hand-to-mouth existence. Too proud to accept the charity of friends, he frequently went without eating and made his bed on park benches. How he spent the next few months of waiting, no one knows. It is almost certain, however, that he threw himself into political work, attending meetings of Republican groups, participating in riots and disturbances, and generally taking an active part in the Republican movement. It is also certain that under the prodding of his only friend from school days, Auguste Chevalier, he wrote another paper and submitted it to the French Academy.

In May he attended a banquet in honor of a group of Republicans who had recently been acquitted of a charge of conspiracy against the king. At the banquet—which was attended by government spies as well as antigovernment Republicans—Galois made an unfortunate toast to the death of the king. For this display of impetuous idealism and lack of judgment, he was arrested the next day at his mother's house, imprisoned, and tried for having provoked an attempt against the life of King Louis Philippe. Despite Galois' seeming determination to make a martyr of himself—he had once said, "If the people need a corpse to be stirred to revolt, I will give them mine"—he was acquitted by the pro-Republican jury. In the eyes of the government, however, he was still a dangerous man.

His freedom lasted a month. In June he was arrested again and held for two months while the police combed the legal code for a crime with which to charge him. They eventually indicted him for wearing an illegal uniform, that of the Artillery Guard, which had been dissolved after the December riots. Galois lingered in prison for three more months awaiting trial. Speedy justice, one of the earmarks of a free country, was absent in France.

When his trial came up at the end of October, he was found guilty and sentenced to six months, an unusually severe penalty for so minor a crime. Furthermore, the five months he had already served while awaiting trial were not subtracted from his sentence. His sister said that he looked fifty years old as he was led away under guard. He had turned twenty just a few days before.

While in prison, Galois received a letter from the Academy, enclosed with which was the manuscript he had sent them nine months before. The Academy was sorry, but they did not understand what M. Galois was talking about. "It is not even possible for us," they wrote, "to give an idea of this paper. M. Galois' proofs are neither sufficiently clear nor sufficiently developed to allow us to judge."

Galois decided to have nothing more to do with the Academy. Instead he would write out his theories and have them published privately. So while still in prison, he began his papers. The introduction is filled with all the bitterness of a man denied three times. "To the men important in science," he wrote, "I owe the fact the first of the two papers in this work is appearing so late. I owe to the men important in the world the fact that I wrote the entire thing in prison." Not only did he rail against the Academy and the government, he bore still another grudge which he felt his paper would settle: "I shall especially have to bear the wild laughter of the École Polytechnique examiners who, having monopolized the printing of mathematical textbooks, will raise their eyebrows because a young man twice refused by them has the pretension to write not a textbook, but a treatise . . . All the foregoing I mention to prove that it is knowingly that I expose myself to the derision of fools."

The fools for whom his papers were intended never got to read them, for Galois never finished them. He spent his rage writing the introduction and after that his enthusiasm waned.

While Galois was in prison, the same cholera epidemic that had threatened Lobatchevsky in Kazan raged through Paris. Afraid that more Republican riots would spring up if their most important political prisoner died, the police transferred Galois to a nursing home. There conditions were much better and restrictions were few. He could even receive guests on weekends. One day a pretty young girl came to visit one of the other prisoners. Évariste became acquainted with her and for the first time in his life fell in love.

At the end of April, 1832, he was freed. But again his freedom was short—a month of freedom, love, and happiness wedged between prison, hatred, bitterness, anguish, and despair. His sweetheart left him after two or three weeks. Évariste wrote to his best friend, Auguste Chevalier, telling him of his heartbreak: "How can I console myself when I have exhausted in one month the greatest source of happiness a man can have? . . ." He became almost psychopathic in his hatred for a world that had treated him so badly: "Pity, never! Hatred, that is all . . . I approve of violence—if not with my mind, then with my heart. I wish to avenge myself for all my sufferings . . . I am disenchanted with everything, even love of glory. How can a world which I detest soil me? Think about it!"

A few days later he was challenged to a duel. Dueling in France was forbidden by law and therefore had to be conducted in secrecy. The mystery surrounding Évariste's duel has never been completely penetrated, but it is certain that the duel was connected with his love affair. The night before he was to meet his opponent, Évariste wrote to his Republican friends saying, "I have been challenged by two patriots and it is impossible for me to refuse. I beg your forgiveness for keeping this from you, but my opponents put me on my honor not to inform any patriot. All I want you to do is to let it be known that I am fighting against my will after having exhausted all means of reconciliation." In another letter he said, "I die the victim of an infamous coquette. My life is snuffed out in a miserable piece of slander. Oh! Why die for so trivial a thing, for something so despicable!"

One plausible reconstruction of the events leading to the duel is that Évariste was misused—or felt he had been—by his sweet-heart. Apparently he called her the kind of names no lady likes to hear, even if she is no lady. Not only did Galois tell her, but her friends as well, just what he thought. This "baleful truth" told to "men who were so little able to listen to it coolly" provoked the duel. Galois did not want to fight. He was sure that he would be killed, so sure that he signed his letters, "I die your friend, E. Galois."

Then, after he said good-by to his friends in a few letters, he proceeded to cram a lifetime's thought into one night, writing out his mathematical theories. Feverishly working all night long under tremendous pressure, he wrote down his proofs. These would be his last hours on earth and he wanted to leave his most important thing behind—his mathematical genius. Death, he was sure, waited for him the next day; he had only this one night left. Yet it was impossible to develop all his theories completely. Thousands of pages have been taken to elaborate on only a por-tion of his work—and how many pages could one man write in a single night? "I have no time," he hastily scrawled again and again in the margins next to uncompleted proofs.

Dawn came and an exhausted Évariste rode off. It was Wednes-day morning, May 30. He was twenty years old and had left a life's work of sixty pages behind. A few hours later a peasant found him shot through the stomach, alone, and lying in a ditch where his seconds had deserted him. Évariste was taken to a hos-pital. He refused the services of a priest and asked only to see his brother. Alfred was sent for, and at the sight of the dying boy, burst into tears. "Don't cry," Galois whispered, "I need all my courage to die at twenty." With his last breath he told Alfred that the police had been his attackers; the police had killed him. Whether this was the truth or merely the ravings of a dying boy who even in his most lucid moments suffered from a persecution complex is not known. Dumas, in his *Mémoires*, states that Galois was killed by Pecheux d'Herbinville, a "delightful young man," in a duel of honor. The Prefect of Police recorded: "M. Galois, a fierce Republican, was killed in a duel by one of his friends." Leopold Infeld, author of Galois' biography *Whom the Gods Love,* inclines toward the belief that the duel and all that led up to it—including Galois falling in love—had been arranged by the police. On Thursday, May 31, 1832, Galois died and was buried in a common burial ground; the exact spot of his grave is unknown. Three thousand people turned out for his funeral. They came to honor Galois, the Republican martyr. History honors Galois the mathematician.

His brother Alfred and his friend, Auguste Chevalier, had been entrusted with his papers. It was up to them to see that they were published and the theories made known. Évariste had naïvely suggested that they "ask Gauss or Jacobi to give their opinion publicly." Gauss, of all people, with his notorious indifference to the work of others, was the last person to ask.

After fourteen years, the papers were published by the French mathematician and editor Liouville. Alfred lived to see his brother's fame spread; and by the end of the nineteenth century, Évariste Galois was considered to be one of the greatest mathematicians of his time. His old headmaster, too, lived to see the name of Galois in the honor roll of mathematics. Asked to comment on his former pupil, he shook his head as though baffled and muttered, "A very strange boy, very strange indeed."

It is impossible to give here more than a sketch of Galois' most important work. He succeeded in proving that equations above the fourth degree cannot generally be solved by radicals and then went on to show what are the necessary conditions for such a solution. This proof, of course, does not rule out the possibility of solving these higher-degree equations by other means. In 1858 two men, Hermite and Kronecker, managed to find nonalgebraic solutions to fifth degree equations.

The study of equations led Galois to originate a new theory known today as the Galois theory of groups. Instead of performing operations on numbers, group theory operates on operations. A complete explanation of the subject is beyond the scope of this book, but there is no overlooking the importance of group theory to modern mathematics. In its concern with the underlying structure of equations, it lays the groundwork for today's higher algebra.

A few days before Galois died he had written, "There are those who are destined to do good but never to experience it. I believe I am one of them." History has proved him right. George Sarton, the late historian of science, declared that "when mathematicians of the future contemplate the personality of Galois at the distance of a few centuries, it will appear to them to be surrounded by the same halo of wonder as those of Euclid, Archimedes, Descartes and Newton."

Georg Cantor
1845-1918

There are times in history—the history of a man as well as a civilization—when one can look back and say, "So this is where it has all been leading. It seems so obvious now, why didn't I realize it before?" A man or a civilization comes to the end of a road; the journey is over; all the wanderings and travels down dead ends and over highways have led to this particular place and suddenly he realizes that he is at the end, the trip is over, the journey completed. Such was the feeling mathematicians had after Georg Cantor guided them over the last stretch of land.

They could rest. Their doubts and fears and wonders of where the road would lead were satisfied. But another road stretched out before them, an ill-defined, treacherous-looking path that both repelled and beguiled—and soon another journey began. The end of one trip suddenly became the beginning of another. Such, too, was the feeling mathematicians had after Georg Cantor opened their eyes to a new and foreign world.

The non-Euclidean revolution had barely begun when Cantor was born in St. Petersburg on March 3, 1845. His father, Georg Woldeman Cantor, a Danish merchant who had settled in Russia, was a Jew who had been converted to Protestantism and rounded out his cosmopolitan religious experiences by marrying a Catholic, Maria Böhm.

From their mother the Cantor children—Georg, his younger brother and sister, Constantin and Sophie—inherited a strong artistic streak which Georg was to express in mathematics, his brother in music, and his sister in painting.

The children were taught at home and later sent to elementary school. Georg, at a very early age, showed enough talent in mathematics for his father to decide he would be a great success as an engineer, a choice as inspired as if Rubens had decided to become a sign painter.

When Georg was eleven, his father began to suffer from the harsh Russian climate, so the family moved to Frankfort, Germany. The children were enrolled in private schools and at fifteen Georg was sent to the Grand-Ducal Higher Polytechnic at Darmstadt to study engineering. Too young and too helpless to assert his own preferences, he submitted to paternal pressure without even a flicker of resentment. Some degree of the nature and subtlety of that pressure shows through in a letter received from his father during the first year: "Your father, or rather your parents, and all the rest of the family in Germany as well as Russia and Denmark, have their eyes on you as the eldest and expect nothing less than [that you become] a shining star on the horizon of engineers."

But two years away from home and the direct influence of his father, two years spent sampling his future profession, were enough to give Georg courage to plead his case. He begged to be allowed to become a mathematician, and finally, just before he was graduated, his father gave in. Georg accepted victory as meekly as, two years earlier, he had accepted defeat. "My dear Papa!" he wrote, "you can just imagine how much your letter pleased me. It settles my future. My last days here have been filled with doubt and indecision. I was torn between duty and desire. How happy I am now that I see it will not displease you if I follow my feeling in my choice. I hope you will live to find pleasure in me, dear father, for my soul, my whole being, lives in my calling; a man can succeed in what he wants to do and toward what he is driven by a secret, unknown voice."

And so, in 1862 Georg went to Zürich to continue his education but left the following year because of his father's death.

In the fall he entered the University of Berlin to study mathematics, physics, and philosophy. Three of Germany's greatest mathematicians, Kummer, Weierstrass, and Kronecker were on the staff. Classes were small, students were few, and instruction was often on an individual basis. Weierstrass was undoubtedly the most influential of the three in shaping Cantor's mathematical ideas and had already made notable progress on the work that his pupil was eventually to perfect: the arithmetic continuum. Kronecker, on the other hand, had no sympathy for what interested his colleague and student. His ideas were the antithesis of Cantor's. The only thing they had in common was genius.

After passing his orals *magna cum laude* in 1867, Cantor received his doctorate. Because no decent teaching positions were available, he resigned himself to the dreary task of instructing young girls in mathematics at a private school. Two years later he joined the faculty of Halle University, a small, second-rate institution.

But even at a second- or third-rate place, a man may have first-rate ideas, and Cantor was such a man.

To understand what problems concerned him and why, it is necessary to take a brief backward glance. The net effect of the non-Euclidean revolution had been to topple mathematics from the throne of Truth. The revolution had been brewing even before the days of Euler. It was now complete. The inevitable scramble for power that follows any revolution began. Religion reasserted her claims that had for so long been drowned out by the creaking and grinding of the universal machine. If science and mathematics rest ultimately on unprovable and possibly erroneous axioms, why should men have more faith in the truth of these axioms than in God? Why should not a faith in God be just as valid as a faith in science?

Philosophies of irrationalism were offered to relieve the scientific disillusion by providing other illusions. Strangely enough, these irrational doctrines held the greatest appeal for and were, with few exceptions, conceived by the very people whose faith in science had been the strongest—the Germans. And ironically, it was two Germans, Gauss and Riemann, and two men schooled by Germans, Lobatchevsky and Bolyai, who had pioneered in non-Euclidean geometry and thus had brought the scientific crisis to a head. By the same token, it was also the German mathematicians—Weierstrass, Dedekind and Cantor—who made the greatest strides in salvaging and rebuilding mathematics.

Within mathematics, the effect of non-Euclidean geometry was to discredit the old idea that mathematical truth is the same as objective, absolute reality. Men now began to realize that there are many mathematical systems—all equally true—built on different sets of postulates or axioms. The axioms may differ, but the rules of logic are the same for each system. Thus, mathematics can be viewed, according to David Hilbert, one of the greatest mathematicians of the twentieth century, as simply a meaningless game with no connection at all with the real world. And yet, as Einstein asked, "How can it be that mathematics, being after all a product of human thought independent of experience, is so admirably adapted to the objects of reality?" How can it be that "meaningless" truths can be so useful? Is it mere coincidence that conic sections describe the orbits of planets; that imaginary numbers describe alternating currents; that the classic Fibonacci series ($\frac{1}{1}$, $\frac{1}{2}$, $\frac{2}{3}$, $\frac{3}{5}$, $\frac{5}{8}$, $\frac{8}{13}$, . . .) describes the arrangement of scales on a pine cone and seeds in a sun-flower; that geometric progressions describe the whorls of a sea-shell; or that π should unexpectedly crop up in the laws of chance? Is it mere coincidence that reason is so useful in probing reality? No one knows.

Despite the marvels of application, mathematics is not claimed to be the same as reality. It is simply a marvelous mental brew. "So far as the theorems of mathematics are about reality, they are not certain; so far as they are certain, they are not about reality," said Einstein, and this seems to be the view of most modern scientists.

Mathematics is a tool capable of carving many models, the best of which is then selected to describe or picture observed facts. But the facts—or the reality behind the facts—and the model do not necessarily correspond exactly—or even at all. As long as the model gives good results, however, it can be called true or at least highly probable.

The question now centers, as noted in the chapter on Lobatchevsky, on what kind of tool or game mathematics is. Is it, as the formalists maintain, purely mental and therefore perfectly rational and logical; or is it imperfect, made up partly of intuition and experience, and therefore, not always reliable?

Cantor belonged to the formalist school rather than to the intuitionist. Like his old teacher, Weierstrass, he believed that there was nothing really wrong with mathematics that a little clear thought could not mend. The flaws were simply due to fuzzy thinking, bad logic, and unwarranted assumptions. He set out to purge mathematics of these errors, to hone the edge of the tool.

A branch of mathematics known as axiomatics developed from these efforts to make mathematics more rigorous. This is the branch that prescribes the methods and techniques of establishing consistent, complete systems, each based on a set of independent and noncontradictory axioms. The basic idea is nothing new. It is older than Euclid. But with the pioneering work of Cantor and several other men, axiomatics became a refined discipline that attempted to formalize the whole body of mathematics, not just the geometric limb.

Initially axiomatics served to put—or to try to put—the existing mathematics on a firm, logical basis. Gradually its operations were extended to formalizing a vast array of mathematical systems based on many different sets of postulates or axioms. Released from its prison of absolute truth, mathematics was free to roam and to develop any kind of systems it wanted—with the one restriction that these systems must be internally consistent. Whether they are useful in describing the world does not matter, for modern mathematicians are no longer concerned with depicting what they see in the most realistic way possible. Like modern painters, they have become more interested in the techniques and methods of their art than in mere description. If their creations happen to have a practical value, so much the better—but the application is only incidental to the creation.

Symbolic logic, or the use of symbols rather than words in proving theorems, was invented so as to enable mathematicians to handle their strange new creations and to free their proofs from the distracting, error-producing meanings that are present when words are used. Although Leibniz had started a completely symbolic system as early as the seventeenth century, George Boole is generally credited with originating symbolic logic and thus providing mathematics with a means to purify itself. Bertrand Russell, one of the staunchest supporters of symbolic logic, wrote, "Pure mathematics was discovered by Boole, in a work which he called *The Laws of Thought* (1854) . . . His book was in fact concerned with formal logic, and this is the same thing as mathematics.

"Pure mathematics consists entirely of assertions to the effect that if such and such a proposition is true of *anything*, then such and such another proposition is true of that thing. It is essential not to discuss whether the first proposition is really true, and not to mention what the anything is of which it is supposed to be true. Both these points would belong to applied mathematics. . . . Thus mathematics may be defined as the subject in which we never know what we are talking about, nor whether what we are saying is true." *

While Cantor had little regard for a strict symbolic logic, he did use the broader axiomatic method of which symbolic logic is a part. In his endeavor to put mathematics on a sound basis, he turned his attention to one of the most irritating problems of all, that of number.

We can imagine numbers plotted on a line similar to one of the axes of Descartes' coordinate geometry. If all rational numbers are plotted—that is, all integers and fractions—it is impossible to find any gaps in the line. Between any two numbers, for example $\frac{1}{2}$ and $\frac{2}{3}$, there is still another number, $\frac{7}{12}$. Between $\frac{1}{2}$ and $\frac{7}{12}$ there is $1\frac{3}{24}$, and between these still another, and so on ad infinitum. Thus, it would seem that all points are filled. In Cantor's words, a reciprocal or one-to-one correspondence would exist between all the points on the line and all the rational numbers.

This system of numbers—neat and tidy—is what Pythagoras imagined "rules the universe." Yet even Pythagoras knew that the system was incomplete. There are points on the line that are not filled by real numbers, as can easily be seen by laying the hypotenuse of a right triangle whose legs are each 1 on the x axis. The point determined by the length of the hypotenuse, $\sqrt{2}$, has no numerical equivalent in the rational system of numbers. All the points of the line are not filled; there is no reciprocal correspondence between the points and numbers.

The length of the hypotenuse is irrational. To fill the gaps in the line irrational numbers must be admitted to the system. But on what basis, aside from that of convenience and necessity, are they to be included, and will their admission result in filling all the gaps? These are the questions Cantor undertook to answer, and in answering them he completely changed the idea of what a number is.

* Bertrand Russell, *Mysticism and Logic,* published by George Allen & Unwin, Ltd.

His approach to the problem was ridiculously simple; his solution was considered by many as simply ridiculous. He began by counting how many integers, how many rational numbers, and how many real numbers there are. Obviously, he could not count up all the numbers, for there are an infinite number of each kind. Furthermore, Cantor was not interested in finding exactly how many numbers there are; he wanted to know how many of each kind there are in relation to the other kinds. So he proceeded along the lines any child might who had not learned to count. If such a child were given two boxes of marbles and asked which box contained more, he could easily find out by taking a marble at a time out of each box. If one box were emptied before the other, he would know that that box had fewer marbles to begin with. If the boxes became empty at the same time he would conclude that they had contained the same amount.

Cantor did the same thing, but instead of using marbles he used numbers; and instead of using boxes he used what he called sets, or classes. A set is simply a collection of similar things. These collections of things can be apples, marbles, people, lines, points, numbers—anything. Cantor chose to let the things or members of his collections be numbers, all having a property in common. That is, the members of one set were even numbers, another odd numbers, another integers, and so forth. He then proceeded to compare the sizes of these different sets by pairing off their members. If each member of one set could be paired with each member of a second set, the sets could be said to be equal in size.

Taking the set of natural numbers, he paired it with the set of even numbers and found that there are as many integers as there are even integers! That this is so can easily be seen (Diagram 28). For every whole number there is a corresponding even number—its double. Thus, he arrived at the remarkable conclusion that when an infinite number of things is being considered, the whole is not greater than its parts. Any infinite set whose members are some kind of whole number is the same size as the set of all the

Integers	Even Numbers
1	2
2	4
3	6
4	8
5	10
.	.
.	.
.	.

DIAGRAM 28

whole numbers. For example, there are as many squares as there are numbers; as many cubes, as many numbers exactly divisible by 10 or 100 or a trillion, as there are whole numbers (Diagram 29). Indeed, as Cantor found and as should be obvious by now, there is no infinite set that is smaller than the set of whole numbers. As to how many numbers are in this set, Cantor adopted the term aleph sub zero, written $\aleph_0$. Aleph is the first letter of the Hebrew alphabet. To distinguish this new number from the finite numbers, he called it transfinite. $\aleph_0$ is as much a number as is 1 or 36 or any other finite number.

Then he tackled the question of whether other transfinite numbers exist; that is, are there larger infinite sets than the infinite set of whole numbers? There seem to be more rational numbers, which include fractions, than there are whole numbers, yet when

Integers	Squares	Cubes	Numbers divisible by 10	Numbers divisible by 100	Numbers divisible by 1 trillion
1	1	1	10	100	1 trillion
2	4	8	20	200	2 trillion
3	9	27	30	300	3 trillion
4	16	64	40	400	4 trillion
5	25	125	50	500	5 trillion
.	.	.	.	.	.
.	.	.	.	.	.

DIAGRAM 29

Cantor paired the two sets he found them to be equal. There are as many fractions and whole numbers put together as there are whole numbers alone!

Before considering the set of real numbers, which includes irrationals, it is only fair to point out that Cantor had some extra information which many readers do not. He knew that Jacques Liouville in 1844 had proved that there are two kinds of irrational numbers: algebraic and transcendental. An algebraic number is one that arises as the root of an algebraic equation. Since there are an infinite number of algebraic equations, there are an infinite number of roots, both rational and irrational. Yet some numbers can never be roots of an algebraic equation. For instance, it is impossible to formulate an equation that will have π as a root, for π only arises when one uses the infinite processes of analysis, not the finite processes of algebra. Nonalgebraic equations such as exponential, trigonometric, and logarithmic ones do not, as a rule, have roots that are algebraic numbers. Nonalgebraic numbers are called transcendental. The best-known transcendental numbers are π and e, which is used as the base of the natural logarithms and is approximately 2.71828. A diagram of real numbers looks like this:

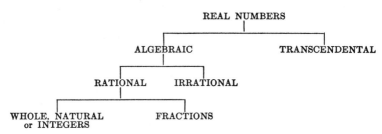

In comparing the size of the set of whole numbers with that of real numbers, Cantor made a distinction between algebraic numbers and the more inclusive real numbers, which includes transcendental as well as algebraic numbers. He first tried pairing the set of whole numbers with the set of algebraic numbers. Through an ingenious method of ordering algebraic equations on the basis of their coefficients and exponents, Cantor was able to show that their roots, i.e., all algebraic numbers, can be paired one-to-one with whole numbers. That is, the set of algebraic numbers is of the same size as the set of whole numbers.

So far the search for a larger infinity has been as rewarding as a dull mystery. All sets seem to be the same size, but Cantor surprised everyone—including himself—when he tried to pair the set of real numbers with the set of whole numbers and found the former to be larger, a great deal larger. The larger size of the set of real numbers is due to the transcendental numbers it contains. When first discovered, transcendentals were thought to be rare, but Cantor showed that exactly the opposite is true. They are not only common, but are more plentiful than any other kind of number.

His proof that the set of real numbers is larger than the set of integers (or the set of rational numbers or algebraic numbers) is quite simple. First he assumed that there is a perfect correspondence between all the integers and all the real numbers from zero to one. (If there is a correspondence between all the integers and all the reals from zero to one, then there is also a correspondence between all the integers and all the reals from zero to infinity.) To effect this correspondence, all the real numbers from zero to one must be listed. Cantor assumed that this list could be made, with the real numbers appearing in the form of infinite decimals, such as:

$$.129578463 \ldots$$
$$.200000000 \ldots$$
$$.298475038 \ldots$$
$$.864703126 \ldots$$
$$.926490003 \ldots$$

Then by a procedure known as the diagonal process, he showed that this list does not contain all the real numbers, that there are other real numbers missing. For instance, a real number different from any number on the list can be formed by choosing as its first digit a figure that differs from the first digit of the first number on the list, a second digit that differs from the second digit of the second number, a third that differs from the third digit of the third number, and so forth. The resulting number then will be different from any other number on the list, for it will differ in at least one digit from every number listed. Since it differs from every number on the list, it is not on the list itself. Therefore, the original assumption that the real numbers can be listed and paired with the integers is false, for it leads to contradictions.

In this way Cantor proved that the set of real numbers is greater than the set of whole numbers. Furthermore, the diagonal process can be used to show that larger and larger infinite sets exist—*that there is no largest infinite set*. Thus, the transfinite numbers, or orders of infinity, like the regular finite numbers, are infinite. It will be recalled that Cantor called the first transfinite number $\aleph_0$. He called the second transfinite number—the one describing the set of all real numbers—C. It has not been proved whether C is the next transfinite number after $\aleph_0$ or whether another number exists between them. It is known, however, that larger transfinite numbers than C exist.

Cantor's theories on infinity provide the solution to some of the oldest problems in mathematics. The paradox of Achilles and the tortoise, for instance, is solved—at least partly. It may be remembered that the problem centered in Achilles' having to run through more points than the tortoise and since each of the racers could be in only one point at a time, Achilles could never catch up with the tortoise. Cantor's proof that the whole of an infinite set—such as points on a line—is not greater than its parts—such as a segment of the line—clears up this aspect of the puzzle. Achilles does not have to run through more points than the tortoise. He has to run through the same number: an infinite number. The question of how either of the racers can run through an infinite number of points in a finite amount of time—or time divided into an infinite number of instants, is solved partly by Cantor's theory of irrationals (to be described) which shows that the sum of an infinite series of a rational number can be a finite number, and partly by Einstein's unification of space and time.

The problem mentioned in the chapter on Descartes about how there can be enough number-points on a segment of the x axis

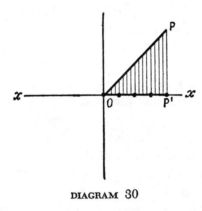

DIAGRAM 30

to account for the number-points on a longer line determined by that segment can now be answered. Both lines, OP and OP′ (Diagram 30) have an infinite number of points, specifically, the transfinite number C. Thus, although one line is longer than the other, it still has the same number of points, for any infinite set of a certain order is equal to any other infinite set of the same order.

Cantor's theory of infinity forms the basis of his theory of irrationals, a theory that for the first time puts irrational numbers on a logical foundation and connects them with the rational number system. Weierstrass, Cantor's teacher at the University of Berlin, had worked on the same theory, and perhaps more than Cantor, deserves credit as the originator.

In dealing with infinity, pre-Cantorians had regarded it as something never reached, an attitude they had in part acquired from the Greeks. Gauss had stated this position perfectly when he said, "I protest against the use of an infinite magnitude as something completed, which is never permissible in mathematics. Infinity is merely a way of speaking, the true meaning being a limit which certain ratios approach indefinitely close." Cantor, however, made a definite distinction between a *potential* and *actual* (completed) infinity, "the former meaning a variable finite magnitude increasing beyond all finite limits, while the latter is a *fixed, constant* magnitude lying beyond all finite magnitudes." This distinction is very important. A potential infinity consists of a process by which a number grows "beyond finite limits." Actual infinity is not a process; it is itself a number.

This same distinction between potential and actual was applied by Cantor to irrational numbers, and finally, to all finite numbers. That is, any finite number can be described either as an infinite process, an endless sort of evolution, or as an actual, fixed constant that represents the completion of the process. The latter view is the traditional one. Yet neither view is new. The former but not the latter has always been applied to infinity; the latter but not the former to rational numbers; and a vague combination of both to irrational numbers. Cantor remedied matters by showing that all numbers, finite and transfinite, can be viewed in both ways.

The two views of infinity look like this: Infinity is the never-reached limit of an infinite number of numbers. That is, the numbers 1, 2, 3, 4, 5, . . . can be continued indefinitely but will never reach their goal of infinity. Viewed in this way, each number in the sequence is only a step in an infinite process. The never-to-be-reached goal or limit, however, can be viewed as a number itself, a transfinite number. This transfinite number is infinity actualized; it is the never-to-be-reached goal when it is reached; it is Cantor's "fixed, constant magnitude lying beyond all finite magnitudes."

In the same way, irrational numbers can be viewed as the goal of an infinite sequence of numbers. The sequence of numbers that have infinity as their goal is easy to visualize. It is the natural sequence, 1, 2, 3, 4, 5, . . . It can also be any sequence generated by the natural sequence, such as 1, 2, 4, 8, 16, . . . , where each successive number is the double of the previous number. As was seen before, these sequences are all the same size: namely $\aleph$, the first cardinal number of the transfinite numbers. The problem now is to represent the infinite sequences that have irrational numbers as their goal. Here, too, Cantor solved the problem in a way so simple that most people would not even think of it. For example, the irrational number $\sqrt{2}$ is an infinite decimal, 1.414214. . . . How can this decimal be represented as an infinite sequence of rational numbers that forever grow larger and larger but that never reach or pass their goal of $\sqrt{2}$? It can be done very simply: 1.4, 1.41, 1.414, 1.4142, 1.41421, 1.414214, . . . Each number in this sequence is only a step in the infinite process of evolving the irrational number, $\sqrt{2}$. Thus, $\sqrt{2}$ is the limit of an infinite process and just as the goal of the natural sequence can be viewed as an actual number, a transfinite number, so the goal of the rational sequence, 1.4, 1.41, 1.414, . . . can also be viewed as a number, an irrational number. This irrational number is now defined, for the first time in history, solely in terms of rational numbers.

Nor do these two views of number apply only to irrational and transfinite numbers. They apply equally to rational numbers. The view of a rational number as an actual entity is well enough known to be omitted. A rational number viewed as the limit of an infinite process is less well known and will be considered.

All rational numbers can be expressed in the form of an infinite decimal. For instance, 4 can be written as 4.0000 . . . , with the zeros going on indefinitely; a fraction such as $\frac{1}{3}$ is .3333333 . . . , and $\frac{1}{7}$ is .1428571428571. . . . Any infinite decimal can, of course, be rewritten as an infinite sequence; and an infinite sequence of this type will, after an infinite number of steps, reach a limiting goal. Thus the rational numbers can also be viewed as the goal

of an infinite process. Viewed in this way, they are no different
from irrational numbers—yet a difference does exist. Whenever
a rational number is expressed as a decimal, the sequence of
digits repeats, and for this reason the decimal is called repeating
or periodic. That is, in the first example (4.0000 . . .) zero repeats itself; in the second (.33333 . . .) three repeats itself; and
in the last example, the sequence .142857 repeats itself. A repeating decimal is characteristic of *all rational* numbers but *never* occurs with *irrationals*. That this is so can easily be seen by taking
the periodic decimal, .142857142857 . . . and setting it equal to
x: $x = .142857.$ . . . (Any operations performed on both sides of
the equation will not change the value of x. That is, the ratio of x
to .142857 . . . will remain the same.) Multiply both sides of the
equation by one million: $1,000,000x = 142,857.142857.$ . . . Now
subtract the first equation from the second:

$$1,000,000x = 142,857.142857...$$
$$- \qquad x = \qquad\quad .142857...$$
$$\overline{}$$
$$999,999x = 142,857.000000...$$

The equation can now be expressed as a terminating fraction:
$x = \dfrac{142,857}{999,999}$. Reduced to its lowest terms, the fraction is $\frac{1}{7}$. Thus,
$\frac{1}{7}$ is the fractional equivalent of the infinite periodic decimal
.142857142857. . . . An irrational number can never be expressed
as a ratio and therefore (as can be seen by working backward)
cannot be a repeating decimal.

These two ways of looking at numbers solved many of the old
problems of mathematics—including that of infinitesimals, which
Berkeley had declared "strained and puzzled . . . our Sense."
Infinitesimals usually have zero as a limiting goal rather than infinity, in which case the sequence shrinks in size (by smaller and
smaller amounts) rather than grows. In both cases, however, the
sequences converge toward limiting goals.

Newton and his followers had only added to the confusion by

using the term *infinitesimal* in a double sense—sometimes meaning the limiting goals ("neither finite quantities, nor quantities infinitely small nor yet nothing") and sometimes meaning a step on the way ("the objects at first fleeting and minute, soon vanishing out of sight"). Actually the problem of infinitesimals was cleared up before Cantor's time. Augustin Cauchy in the 1840's had substituted the concept of limits for infinitesimals. Thus, the formal concept of limits preceded the broad theory of which it is an integral part. It is an example, if a pun may be permitted, of a "limited" application of a larger theory that has not yet been formulated. It is Cantor's theory in miniature.

That the Cauchy theory of limits appeared out of chronological order is not at all surprising. There are many, many examples both in and out of mathematics in which pieces of a larger theory are used consciously or unconsciously before the whole theory has been conceived. For instance, Descartes implicitly used the arithmetic continuum (real-number domain) in his analytic geometry without any clear understanding of this continuum. The paradoxes of Zeno have been interpreted as being a protest against the artificial separation of time and space—and Zeno lived more than two thousand years before Einstein formulated the concept of four-dimensional time-space.

As a matter of fact, Cantor's whole theory of irrationals and transfinites is far from a completely new creation—despite its strangeness and seeming originality. It was all there before; he merely gave the first clear and formal expression to the vague notions that men had tacitly and unwittingly employed for centuries. Far from new, his theories are but the culmination of an historical trend that began with Descartes. Viewed from the vantage point of the twentieth century, it seems almost incredible that no one had formulated them sooner, for they are the basis of the calculus. Newton's indivisibilia or infinitesimals are Cantor's steps in the infinite process of evolving a number. A fluxion or *ultima ratio*—today called derivative or differential—is Cantor's real number, the realized goal or limit of the infinite process. Even the apparently startling conclusion that a part of infinity is as great as the whole can be drawn from the simplest problem in calculus. Either this conclusion must be drawn or else the whole machinery of calculus must be abandoned. For instance, in the problem of the velocity of a falling object considered in the chapter on Newton, it was noted that the numbers denoting time in seconds proceed in a natural, consecutive order: 1, 2, 3, 4, . . . The numbers denoting the corresponding distances, however, proceed in larger and larger leaps: 16, 64, 144, 256, . . . Unless these two sets of numbers can be matched on a one-to-one correspondence, there eventually will not be any numbers left to denote the distance at a certain time. The only solution is to conclude that there are as many numbers in the infinite set, 1, 2, 3, 4, . . . as in the set 16, 64, 144, 256. . . . Cantor, of course, not only drew this apparently illogical conclusion but proved it to be logical. Nor was he the first man to draw the conclusion. As far back as 1636, Galileo had written, "We cannot speak of infinite quantities as being the one greater or less than or equal to the other . . . We must say that there are as many squares as there are numbers, because they are just as numerous as their roots, and all the numbers are roots . . . In an infinite number, if one could conceive of such a thing, he would be forced to admit that there are as many squares as there are numbers all taken together . . . The number of squares is infinite and the number of their roots is infinite; neither is the number of squares less than the totality of all numbers, nor the latter greater than the former." Truly, it seems that every great idea has been thought before, but that men do not always recognize greatness when they see it.

One would think, for instance, that Cantor's theories, which provide the solution to so many long-standing problems, would have been immediately hailed as among the mathematical triumphs of the century, if not of all time. Sad to say, they were not. They were scorned, ridiculed, and considered slightly insane.

Even sadder to say, Cantor eventually changed places with his brainchildren. They were eventually accepted and respected; Cantor, broken by the terrible struggle, went insane.

After three years at the University of Halle, Cantor was made an assistant professor. Two years later, at the age of twenty-nine, he married a young German girl, Vally Guttmann, by whom he had six children—two boys and four girls. He spent an idyllic honeymoon at Interlaken in Switzerland where he met and befriended another young mathematician, Richard Dedekind. Two years earlier, Dedekind had published a paper on his own theory of irrationals. The two men undoubtedly spent many hours discussing their respective theories. But even more important than the exchange of ideas was the encouragement they gave each other, for both were to enjoy a remarkable indifference on the part of the rest of the mathematical world and to reap all the rewards of genius too original for its time: relegation to obscure, underpaid positions.

Cantor published his first major paper on his theory of sets later that year. The paper had been submitted and approved months earlier, but one of the editors of the journal in which it was to appear deliberately delayed publication. The editor was Leopold Kronecker, one of Cantor's old teachers at the University of Berlin.

This delay was not simply a matter of procrastination on Kronecker's part. It was a malicious bit of academic censorship and an even larger bit of professional jealousy.

Kronecker is famous (among other things) for the statement, "God made the integers; the rest is the work of man." He believed this sincerely. Negative numbers, fractions, imaginaries, and especially irrational numbers were anathema to him, the source of all mathematics' troubles; he advocated banishing them completely. As for infinity, it was simply a process and never, never could be regarded as an actual number. A mathematical entity that could not be constructed in a finite number of steps was sheer nonsense in his eyes—nor is he alone even today in condemning the actualization of the infinite. His views, as should be obvious, are diametrically opposed to Cantor's. And Kronecker took advantage of his superior professional position to suppress the Cantorian heresy.

Nor was Cantor the only object of Kronecker's oppression. Another mathematician at the University of Berlin, Weierstrass, a colleague of Kronecker's but an opponent of his ideas, was deeply hurt by these attacks. "I can no longer take the same joy that I used to take in my teaching," he wrote near the end of his career. "Kronecker uses his authority to proclaim that all those who up to now have labored to establish the theory of functions (closely connected with infinity) are sinners before the Lord . . . such a verdict from a man whose eminent talent and distinguished performance in mathematical research I admire as sincerely and with as much pleasure as all his colleagues, is not only humiliating . . . but is a direct appeal to the younger generation to desert their present leaders and to rally around him as the disciple of a new system that is inevitable. Truly it is sad, and it fills me with a bitter grief, to see a man whose glory is without a stain, let himself be driven . . . to utterances whose injurious effect upon others he seems not to realize."

To what extent Kronecker's actions were motivated by intellectual conviction and to what extent by professional jealousy is difficult to assess. Certainly there was some of both. He could make Weierstrass' life miserable, but he could not really harm him professionally—Weierstrass' position was too secure for that. But Cantor was another matter. Not only was he grossly inept at defending himself, but he was only a lowly profesor at an insignificant, backwater university. Although he was promoted to a full professorship in 1879, it was a meaningless promotion with no real prestige attached. Kronecker made it his business to see that Cantor stayed where he was.

Berlin was at that time the center of mathematics, the place from which the new ideas flowed forth, the magnet toward which all rising young mathematicians were drawn. Kronecker did not want the opposition on the faculty of the university strengthened by the addition of any Weierstrass adherents—especially Cantor, who had been banging at the gates for a number of years. All of Cantor's applications for a post at Berlin were turned down again and again through the conniving of Kronecker.

Cantor's belief that all the prejudice against him had been "systematically stimulated by Kronecker" was no neurotic delusion. He was not the only person who noticed the uncalled-for hostility. Minkowski, later Einstein's mathematics teacher, observed—without actually mentioning names—that "it is very much to be regretted that an opposition by a very famous mathematician on grounds not purely mathematical could disturb the joys of Cantor in his scientific researches." Schoenflies was blunter in his accusations, noting that Cantor's theories "had found the sharpest opposition from Kronecker. It is not going beyond permissible limits if I say that the attitude of Kronecker was bound to create the impression that Cantor was, as a person, a mathematician, and as a teacher, a spoiler of the young . . . Cantor was isolated and forsaken even by those whom he respected, indeed, even by Weierstrass."

By 1884, the whole of Cantor's theories had been published and largely ignored, thanks to Kronecker's conniving. One of the few people who did not ignore them was a young Scandinavian by the name of Mittag-Leffler. To him Cantor poured out his troubles, and in one year alone wrote him an average of a letter a week complaining of Kronecker's persecution. But by that time it was too late for help. Kronecker's aggressive attacks had become too much to bear. Cantor had never been what might be called self-assertive. In his relationship with his father, he had submitted dutifully—and now again he was subdued by the overwhelming force of another human being. Like all the vanquished, he paid a terrible toll in self-respect. He became very depressed and lost all faith in himself and in his work. Even the words of Minkowski failed to cheer him: "Future generations will regard Cantor as one of the most deep-thinking mathematicians of these times."

In the spring of 1884, Cantor had a complete nervous breakdown. Both his family and his doctors laid the blame at Kronecker's door. It was felt that if the two men could somehow reconcile their differences, Cantor's mental health would be improved.

A temporary truce was effected, and Cantor's health did improve. By 1885 he was his old self again with one exception. He no longer wrote. His creativity was gone. During the rest of his life he published only three papers.

Kronecker's attacks soon began again, more fierce than ever. He devoted his lectures to ripping apart Cantor's theories, continued his intrigues to keep the poor man buried away at Halle, and eliminated all Cantorian articles from *Crelle's Journal,* of which he was editor.

Then in 1891, Kronecker died and his evil influence gradually disappeared. Slowly Cantor began to receive the recognition he deserved—after having waited more than twenty years. He was made an honorary member of the London Mathematical Society, elected a corresponding member of the Society of Sciences at Göttingen, and in 1904 was awarded a medal by the Royal Society of London.

Remembering his own earlier experiences, Cantor was always ready with a word of praise and encouragement for the men who were still struggling. In addition, he founded a journal, the *Deutsche Mathematiker-Vereinigung,* as a vehicle for the works of young researchers who might not be able to voice themselves in the journals controlled by established mathematicians.

But for him, however, recognition had come too late. The unhappiness and bitterness of so many years could not be wiped out by a sudden spurt of fame and glory. Honored throughout the world, he felt like a stranger in his own country. "The Germans do not know me," he wrote to a friend, "although I have lived among them for fifty-two years." He never received a better appointment than his one at Halle for he was too old and too ill to go anywhere else. His nervous attacks became more frequent and of longer duration until he was forced to resign. At the age of seventy-two, Georg Cantor died in a psychiatric hospital on January 6, 1918.

With Cantor the end of a mathematical road has been reached. Infinity—the whole basis of analysis and irrational numbers—has been put on a sound foundation. The threads of two thousand years have been knit together into one consistent theory. But a new road opens up—there is always, fortunately, a new road.

One of the first obstacles encountered concerns Cantor's set theory and his assertion that there are an infinite number of transfinite numbers, i.e., that there is no last transfinite number. If this is true, it leads to the same sort of paradox that the assertion, "all of men's ideas are determined," leads to. That is, if all infinite sets can be assigned a transfinite number, then there must be a set whose members include all transfinite numbers. This set, then, would itself be a last transfinite number—yet Cantor said there is no last such number. Furthermore, does this set, since it includes all infinite sets, include itself? Whether the answer is yes or no, still another paradox arises. If it includes itself, then all infinite sets can be divided into two kinds, those that have themselves as members and those that do not, which in turn leads to the question that Bertrand Russell asked, whether the set of all sets which are not members of themselves is a member of itself. It is simply futile to pursue these paradoxes any further. They simply put us in the position of Mr. Russell's barber who shaves only and all men in a village who do not shave themselves. Does the barber shave himself?

There are those who say these paradoxes or antinomies arising from Cantor's theory of the infinite can be resolved and those who maintain that we must discard infinity and the infinite processes, for they lead inevitably to paradoxes. "From the paradise created for us by Cantor, no one will drive us out," declared David Hilbert. "We have stormed the heavens," said Weyl from the opposite camp, "but succeeded only in building fog upon fog, a mist which will not support anybody who earnestly desires to stand upon it."

The formalists, who believe that the paradoxes can be eliminated and that mathematics can be made completely logical and rigorous, have a two-thousand-year tradition behind them. The concept of mathematics as the science that studies quantities has changed to "the science which draws necessary conclusions," but like the ancient Greeks, the formalists believe these conclusions proceed from a firm basis of logic and reason. The intuitionists, on the other hand, believe that not only is the basis not firm, but that it is incapable of being made firm. To support their charges they have some indications from quantum physics that logic and the axiomatic method are insufficient in the construction of mathematical models. Furthermore, they have a proof formulated by Kurt Gödel in 1931 which states that it is impossible to prove the internal consistency of a mathematical system that includes cardinal numbers. This proof, often regarded as one of the most important of recent times, does not mean that all systems are inconsistent. It simply means that we can never know if a system is consistent. In the last resort, we must appeal to faith.

In the hands of the formalists and intuitionists, the mathematical tool has become a double-edged sword in a duel that includes not only mathematics, but all of science and philosophy as well. It is too soon to tell which side will win—and perhaps this, too, is one of those questions whose answer we shall never know.

Selected Bibliography

PYTHAGORAS, EUCLID, AND ARCHIMEDES
Cajori, F. C.: *A History of Mathematics,* New York.
Heath, Sir Thomas: *History of Greek Mathematics,* 2 vols., Oxford.

CARDANO
Cardano, Girolamo: *De vita propria liber,* Milan, 1821. Also in English edition: *The Book of My Life,* translated by Jean Stoner, New York, 1930.
Morley, Henry: *Jerome Cardan: The Life of Girolamo Cardano of Milan,* London, 1854.

DESCARTES
Aubrey, John: *Brief Lives,* Oxford, 1898.
Baillet, Adrien: *The Life of Monsieur Des Cartes,* London, 1693.
Descartes, René: *Discourse on Method.*
Haldane, Elizabeth S.: *Descartes, His Life and Times,* London, 1905.

PASCAL
Bishop, Morris G.: *Pascal, The Life of Genius,* New York, 1936.
Pascal, Blaise: *Pensées de Pascal,* Paris, 1848. Also in English edition: *Pascal's Pensées,* translated by W. F. Trotter, New York, 1958.

NEWTON
Biot, Jean Baptiste: *Life of Sir Isaac Newton,* London, 1829.
Brewster, Sir David: *Life of Newton,* London, 1831.
Brodetsky, Selig: *Sir Isaac Newton, A Brief Account of His Life and Work,* London, 1921.

EULER

Langer, Rudolph E.: "The Life of Leonard Euler," *Scripta Mathematica*, Vol. 3, New York, 1935.

GAUSS

Dunnington, Guy Waldo: *Carl Friedrich Gauss: Titan of Science*, New York, 1955.

Teubner, B. G.: *Briefwechsel zwischen Gauss and Bolyai*, Leipzig, 1899.

Worbs, Erich: *Carl Friedrich Gauss: Ein Lebensbild*, Leipzig, 1755.

LOBATCHEVSKY

Engel, F.: *N. I. Lobatschefskij, Zwei geometrische Abhandlungen aus dem Russischen übersetzt mit Anmerkungen und mit einer Biographie des Verfassers*, Leipzig, 1899.

GALOIS

Davidson, Gustav: "The Most Tragic Story in the Annals of Mathematics, The Life of Evariste Galois," *Scripta Mathematica*, Vol. 6, no. 2, June, 1939, New York.

Infeld, Leopold: *Whom the Gods Love*, New York, 1948.

CANTOR

Smith, David E.: *Portraits of Eminent Mathematicians with Brief Biographical Sketches*, New York, 1936-38.

Turnbull, Herbert W.: *The Great Mathematicians*, London, 1929.

GENERAL

D'Abro, A.: *Decline of Mechanism*, New York, 1939.

Dantzig, Tobias: *Number: The Language of Science*, New York, 1930.

Denbow, Carl H., and Goedicke, Victor: *Foundations of Mathematics*, New York, 1959.

Dubisch, Roy: *The Nature of Number*, New York, 1952.

Farrington, Benjamin: *Greek Science*, 2 vols., Harmondsworth, 1944 and 1949.

Heisenberg, Werner: *The Physical Principles of the Quantum Theory*, Chicago, 1930.

Jeans, Sir James: *Physics and Philosophy*, Ann Arbor, 1958.

Kline, Morris: *Mathematics in Western Culture*, New York, 1953.

Newman, James R., Editor: *The World of Mathematics*, 4 vols., New York, 1956.

Russell, Bertrand: *Mysticism and Logic*, London.

Whitehead, Alfred North: *Science and the Modern World*, New York, 1925.

Index

243

A CATALOG OF SELECTED
DOVER BOOKS
IN ALL FIELDS OF INTEREST

A CATALOG OF SELECTED DOVER
BOOKS IN ALL FIELDS OF INTEREST

CONCERNING THE SPIRITUAL IN ART, Wassily Kandinsky. Pioneering work by father of abstract art. Thoughts on color theory, nature of art. Analysis of earlier masters. 12 illustrations. 80pp. of text. 5⅜ × 8½. 23411-8 Pa. $3.95

ANIMALS: 1,419 Copyright-Free Illustrations of Mammals, Birds, Fish, Insects, etc., Jim Harter (ed.). Clear wood engravings present, in extremely lifelike poses, over 1,000 species of animals. One of the most extensive pictorial sourcebooks of its kind. Captions. Index. 284pp. 9 × 12. 23766-4 Pa. $11.95

CELTIC ART: The Methods of Construction, George Bain. Simple geometric techniques for making Celtic interlacements, spirals, Kells-type initials, animals, humans, etc. Over 500 illustrations. 160pp. 9 × 12. (USO) 22923-8 Pa. $9.95

AN ATLAS OF ANATOMY FOR ARTISTS, Fritz Schider. Most thorough reference work on art anatomy in the world. Hundreds of illustrations, including selections from works by Vesalius, Leonardo, Goya, Ingres, Michelangelo, others. 593 illustrations. 192pp. 7⅛ × 10¼. 20241-0 Pa. $8.95

CELTIC HAND STROKE-BY-STROKE (Irish Half-Uncial from "The Book of Kells"): An Arthur Baker Calligraphy Manual, Arthur Baker. Complete guide to creating each letter of the alphabet in distinctive Celtic manner. Covers hand position, strokes, pens, inks, paper, more. Illustrated. 48pp. 8¼ × 11.
24336-2 Pa. $3.95

EASY ORIGAMI, John Montroll. Charming collection of 32 projects (hat, cup, pelican, piano, swan, many more) specially designed for the novice origami hobbyist. Clearly illustrated easy-to-follow instructions insure that even beginning papercrafters will achieve successful results. 48pp. 8¼ × 11. 27298-2 Pa. $2.95

THE COMPLETE BOOK OF BIRDHOUSE CONSTRUCTION FOR WOOD-WORKERS, Scott D. Campbell. Detailed instructions, illustrations, tables. Also data on bird habitat and instinct patterns. Bibliography. 3 tables. 63 illustrations in 15 figures. 48pp. 5¼ × 8½. 24407-5 Pa. $1.95

BLOOMINGDALE'S ILLUSTRATED 1886 CATALOG: Fashions, Dry Goods and Housewares, Bloomingdale Brothers. Famed merchants' extremely rare catalog depicting about 1,700 products: clothing, housewares, firearms, dry goods, jewelry, more. Invaluable for dating, identifying vintage items. Also, copyright-free graphics for artists, designers. Co-published with Henry Ford Museum & Greenfield Village. 160pp. 8¼ × 11. 25780-0 Pa. $9.95

HISTORIC COSTUME IN PICTURES, Braun & Schneider. Over 1,450 costumed figures in clearly detailed engravings—from dawn of civilization to end of 19th century. Captions. Many folk costumes. 256pp. 8⅜ × 11¾. 23150-X Pa. $11.95

CATALOG OF DOVER BOOKS

STICKLEY CRAFTSMAN FURNITURE CATALOGS, Gustav Stickley and L. &
J. G. Stickley. Beautiful, functional furniture in two authentic catalogs from 1910.
594 illustrations, including 277 photos, show settles, rockers, armchairs, reclining
chairs, bookcases, desks, tables. 183pp. 6½ × 9¼.   23838-5 Pa. $8.95

AMERICAN LOCOMOTIVES IN HISTORIC PHOTOGRAPHS: 1858 to 1949,
Ron Ziel (ed.). A rare collection of 126 meticulously detailed official photographs,
called "builder portraits," of American locomotives that majestically chronicle the
rise of steam locomotive power in America. Introduction. Detailed captions. xi +
129pp. 9 × 12.   27393-8 Pa. $12.95

AMERICA'S LIGHTHOUSES: An Illustrated History, Francis Ross Holland, Jr.
Delightfully written, profusely illustrated fact-filled survey of over 200 American
lighthouses since 1716. History, anecdotes, technological advances, more. 240pp.
8 × 10¾.   25576-X Pa. $11.95

TOWARDS A NEW ARCHITECTURE, Le Corbusier. Pioneering manifesto by
founder of "International School." Technical and aesthetic theories, views of
industry, economics, relation of form to function, "mass-production split" and
much more. Profusely illustrated. 320pp. 6⅛ × 9¼. (USO)   25023-7 Pa. $8.95

HOW THE OTHER HALF LIVES, Jacob Riis. Famous journalistic record,
exposing poverty and degradation of New York slums around 1900, by major social
reformer. 100 striking and influential photographs. 233pp. 10 × 7⅞.
          22012-5 Pa $10.95

FRUIT KEY AND TWIG KEY TO TREES AND SHRUBS, William M. Harlow.
One of the handiest and most widely used identification aids. Fruit key covers 120
deciduous and evergreen species; twig key 160 deciduous species. Easily used. Over
300 photographs. 126pp. 5⅜ × 8½.   20511-8 Pa. $3.95

COMMON BIRD SONGS, Dr. Donald J. Borror. Songs of 60 most common U.S.
birds: robins, sparrows, cardinals, bluejays, finches, more—arranged in order of
increasing complexity. Up to 9 variations of songs of each species.
          Cassette and manual 99911-4 $8.95

ORCHIDS AS HOUSE PLANTS, Rebecca Tyson Northen. Grow cattleyas and
many other kinds of orchids—in a window, in a case, or under artificial light. 63
illustrations. 148pp. 5⅜ × 8½.   23261-1 Pa. $3.95

MONSTER MAZES, Dave Phillips. Masterful mazes at four levels of difficulty.
Avoid deadly perils and evil creatures to find magical treasures. Solutions for all 32
exciting illustrated puzzles. 48pp. 8¼ × 11.   26005-4 Pa. $2.95

MOZART'S DON GIOVANNI (DOVER OPERA LIBRETTO SERIES), Wolfgang Amadeus Mozart. Introduced and translated by Ellen H. Bleiler. Standard
Italian libretto, with complete English translation. Convenient and thoroughly
portable—an ideal companion for reading along with a recording or the performance itself. Introduction. List of characters. Plot summary. 121pp. 5¼ × 8½.
          24944-1 Pa. $2.95

TECHNICAL MANUAL AND DICTIONARY OF CLASSICAL BALLET, Gail
Grant. Defines, explains, comments on steps, movements, poses and concepts.
15-page pictorial section. Basic book for student, viewer. 127pp. 5⅜ × 8½.
          21843-0 Pa. $3.95

CATALOG OF DOVER BOOKS

BRASS INSTRUMENTS: Their History and Development, Anthony Baines. Authoritative, updated survey of the evolution of trumpets, trombones, bugles, cornets, French horns, tubas and other brass wind instruments. Over 140 illustrations and 48 music examples. Corrected and updated by author. New preface. Bibliography. 320pp. $5\frac{3}{8} \times 8\frac{1}{2}$. 27574-4 Pa. $9.95

HOLLYWOOD GLAMOR PORTRAITS, John Kobal (ed.). 145 photos from 1926–49. Harlow, Gable, Bogart, Bacall; 94 stars in all. Full background on photographers, technical aspects. 160pp. $8\frac{3}{8} \times 11\frac{1}{4}$. 23352-9 Pa. $11.95

MAX AND MORITZ, Wilhelm Busch. Great humor classic in both German and English. Also 10 other works: "Cat and Mouse," "Plisch and Plumm," etc. 216pp. $5\frac{3}{8} \times 8\frac{1}{2}$. 20181-3 Pa. $5.95

THE RAVEN AND OTHER FAVORITE POEMS, Edgar Allan Poe. Over 40 of the author's most memorable poems: "The Bells," "Ulalume," "Israfel," "To Helen," "The Conqueror Worm," "Eldorado," "Annabel Lee," many more. Alphabetic lists of titles and first lines. 64pp. $5\frac{3}{16} \times 8\frac{1}{4}$. 26685-0 Pa. $1.00

SEVEN SCIENCE FICTION NOVELS, H. G. Wells. The standard collection of the great novels. Complete, unabridged. First Men in the Moon, Island of Dr. Moreau, War of the Worlds, Food of the Gods, Invisible Man, Time Machine, In the Days of the Comet. Total of 1,015pp. $5\frac{3}{8} \times 8\frac{1}{2}$. (USO) 20264-X Clothbd. $29.95

AMULETS AND SUPERSTITIONS, E. A. Wallis Budge. Comprehensive discourse on origin, powers of amulets in many ancient cultures: Arab, Persian, Babylonian, Assyrian, Egyptian, Gnostic, Hebrew, Phoenician, Syriac, etc. Covers cross, swastika, crucifix, seals, rings, stones, etc. 584pp. $5\frac{3}{8} \times 8\frac{1}{2}$. 23573-4 Pa. $12.95

RUSSIAN STORIES/PYCCKNE PACCKA3bl: A Dual-Language Book, edited by Gleb Struve. Twelve tales by such masters as Chekhov, Tolstoy, Dostoevsky, Pushkin, others. Excellent word-for-word English translations on facing pages, plus teaching and study aids, Russian/English vocabulary, biographical/critical introductions, more. 416pp. $5\frac{3}{8} \times 8\frac{1}{2}$. 26244-8 Pa. $8.95

PHILADELPHIA THEN AND NOW: 60 Sites Photographed in the Past and Present, Kenneth Finkel and Susan Oyama. Rare photographs of City Hall, Logan Square, Independence Hall, Betsy Ross House, other landmarks juxtaposed with contemporary views. Captures changing face of historic city. Introduction. Captions. 128pp. $8\frac{1}{4} \times 11$. 25790-8 Pa. $9.95

AIA ARCHITECTURAL GUIDE TO NASSAU AND SUFFOLK COUNTIES, LONG ISLAND, The American Institute of Architects, Long Island Chapter, and the Society for the Preservation of Long Island Antiquities. Comprehensive, well-researched and generously illustrated volume brings to life over three centuries of Long Island's great architectural heritage. More than 240 photographs with authoritative, extensively detailed captions. 176pp. $8\frac{1}{4} \times 11$. 26946-9 Pa. $14.95

NORTH AMERICAN INDIAN LIFE: Customs and Traditions of 23 Tribes, Elsie Clews Parsons (ed.). 27 fictionalized essays by noted anthropologists examine religion, customs, government, additional facets of life among the Winnebago, Crow, Zuni, Eskimo, other tribes. 480pp. $6\frac{1}{8} \times 9\frac{1}{4}$. 27377-6 Pa. $10.95

FRANK LLOYD WRIGHT'S HOLLYHOCK HOUSE, Donald Hoffmann. Lavishly illustrated, carefully documented study of one of Wright's most controversial residential designs. Over 120 photographs, floor plans, elevations, etc. Detailed perceptive text by noted Wright scholar. Index. 128pp. 9¼ × 10¾.
27133-1 Pa. $11.95

THE MALE AND FEMALE FIGURE IN MOTION: 60 Classic Photographic Sequences, Eadweard Muybridge. 60 true-action photographs of men and women walking, running, climbing, bending, turning, etc., reproduced from rare 19th-century masterpiece. vi + 121pp. 9 × 12. 24745-7 Pa. $10.95

1001 QUESTIONS ANSWERED ABOUT THE SEASHORE, N. J. Berrill and Jacquelyn Berrill. Queries answered about dolphins, sea snails, sponges, starfish, fishes, shore birds, many others. Covers appearance, breeding, growth, feeding, much more. 305pp. 5¼ × 8¼. 23366-9 Pa. $7.95

GUIDE TO OWL WATCHING IN NORTH AMERICA, Donald S. Heintzelman. Superb guide offers complete data and descriptions of 19 species: barn owl, screech owl, snowy owl, many more. Expert coverage of owl-watching equipment, conservation, migrations and invasions, etc. Guide to observing sites. 84 illustrations. xiii + 193pp. 5⅜ × 8½. 27344-X Pa. $7.95

MEDICINAL AND OTHER USES OF NORTH AMERICAN PLANTS: A Historical Survey with Special Reference to the Eastern Indian Tribes, Charlotte Erichsen-Brown. Chronological historical citations document 500 years of usage of plants, trees, shrubs native to eastern Canada, northeastern U.S. Also complete identifying information. 343 illustrations. 544pp. 6½ × 9¼. 25951-X Pa. $12.95

STORYBOOK MAZES, Dave Phillips. 23 stories and mazes on two-page spreads: Wizard of Oz, Treasure Island, Robin Hood, etc. Solutions. 64pp. 8¼ × 11.
23628-5 Pa. $2.95

NEGRO FOLK MUSIC, U.S.A., Harold Courlander. Noted folklorist's scholarly yet readable analysis of rich and varied musical tradition. Includes authentic versions of over 40 folk songs. Valuable bibliography and discography. xi + 324pp. 5⅜ × 8½. 27350-4 Pa. $7.95

MOVIE-STAR PORTRAITS OF THE FORTIES, John Kobal (ed.). 163 glamor, studio photos of 106 stars of the 1940s: Rita Hayworth, Ava Gardner, Marlon Brando, Clark Gable, many more. 176pp. 8⅜ × 11¼. 23546-7 Pa. $10.95

BENCHLEY LOST AND FOUND, Robert Benchley. Finest humor from early 30s, about pet peeves, child psychologists, post office and others. Mostly unavailable elsewhere. 73 illustrations by Peter Arno and others. 183pp. 5⅜ × 8½.
22410-4 Pa. $5.95

YEKL and THE IMPORTED BRIDEGROOM AND OTHER STORIES OF YIDDISH NEW YORK, Abraham Cahan. Film Hester Street based on Yekl (1896). Novel, other stories among first about Jewish immigrants on N.Y.'s East Side. 240pp. 5⅜ × 8½. 22427-9 Pa. $6.95

SELECTED POEMS, Walt Whitman. Generous sampling from *Leaves of Grass*. Twenty-four poems include "I Hear America Singing," "Song of the Open Road," "I Sing the Body Electric," "When Lilacs Last in the Dooryard Bloom'd," "O Captain! My Captain!"—all reprinted from an authoritative edition. Lists of titles and first lines. 128pp. 5³⁄₁₆ × 8¼. 26878-0 Pa. $1.00

THE BEST TALES OF HOFFMANN, E. T. A. Hoffmann. 10 of Hoffmann's most important stories: "Nutcracker and the King of Mice," "The Golden Flowerpot," etc. 458pp. 5⅜ × 8½. 21793-0 Pa. $8.95

FROM FETISH TO GOD IN ANCIENT EGYPT, E. A. Wallis Budge. Rich detailed survey of Egyptian conception of "God" and gods, magic, cult of animals, Osiris, more. Also, superb English translations of hymns and legends. 240 illustrations. 545pp. 5⅜ × 8½. 25803-3 Pa. $11.95

FRENCH STORIES/CONTES FRANÇAIS: A Dual-Language Book, Wallace Fowlie. Ten stories by French masters, Voltaire to Camus: "Micromegas" by Voltaire; "The Atheist's Mass" by Balzac; "Minuet" by de Maupassant; "The Guest" by Camus, six more. Excellent English translations on facing pages. Also French-English vocabulary list, exercises, more. 352pp. 5⅜ × 8½. 26443-2 Pa. $8.95

CHICAGO AT THE TURN OF THE CENTURY IN PHOTOGRAPHS: 122 Historic Views from the Collections of the Chicago Historical Society, Larry A. Viskochil. Rare large-format prints offer detailed views of City Hall, State Street, the Loop, Hull House, Union Station, many other landmarks, circa 1904–1913. Introduction. Captions. Maps. 144pp. 9⅜ × 12¼. 24656-6 Pa. $12.95

OLD BROOKLYN IN EARLY PHOTOGRAPHS, 1865–1929, William Lee Younger. Luna Park, Gravesend race track, construction of Grand Army Plaza, moving of Hotel Brighton, etc. 157 previously unpublished photographs. 165pp. 8⅞ × 11¾. 23587-4 Pa. $13.95

THE MYTHS OF THE NORTH AMERICAN INDIANS, Lewis Spence. Rich anthology of the myths and legends of the Algonquins, Iroquois, Pawnees and Sioux, prefaced by an extensive historical and ethnological commentary. 36 illustrations. 480pp. 5⅜ × 8½. 25967-6 Pa. $8.95

AN ENCYCLOPEDIA OF BATTLES: Accounts of Over 1,560 Battles from 1479 B.C. to the Present, David Eggenberger. Essential details of every major battle in recorded history from the first battle of Megiddo in 1479 B.C. to Grenada in 1984. List of Battle Maps. New Appendix covering the years 1967–1984. Index. 99 illustrations. 544pp. 6½ × 9¼. 24913-1 Pa. $14.95

SAILING ALONE AROUND THE WORLD, Captain Joshua Slocum. First man to sail around the world, alone, in small boat. One of great feats of seamanship told in delightful manner. 67 illustrations. 294pp. 5⅜ × 8½. 20326-3 Pa. $5.95

ANARCHISM AND OTHER ESSAYS, Emma Goldman. Powerful, penetrating, prophetic essays on direct action, role of minorities, prison reform, puritan hypocrisy, violence, etc. 271pp. 5⅜ × 8½. 22484-8 Pa. $5.95

MYTHS OF THE HINDUS AND BUDDHISTS, Ananda K. Coomaraswamy and Sister Nivedita. Great stories of the epics; deeds of Krishna, Shiva, taken from puranas, Vedas, folk tales; etc. 32 illustrations. 400pp. 5⅜ × 8½. 21759-0 Pa. $9.95

BEYOND PSYCHOLOGY, Otto Rank. Fear of death, desire of immortality, nature of sexuality, social organization, creativity, according to Rankian system. 291pp. 5⅜ × 8½. 20485-5 Pa. $7.95

A THEOLOGICO-POLITICAL TREATISE, Benedict Spinoza. Also contains unfinished Political Treatise. Great classic on religious liberty, theory of government on common consent. R. Elwes translation. Total of 421pp. 5⅜ × 8½. 20249-6 Pa. $8.95

MY BONDAGE AND MY FREEDOM, Frederick Douglass. Born a slave, Douglass became outspoken force in antislavery movement. The best of Douglass' auto-biographies. Graphic description of slave life. 464pp. 5⅜ × 8½.   22457-0 Pa. $8.95

FOLLOWING THE EQUATOR: A Journey Around the World, Mark Twain. Fascinating humorous account of 1897 voyage to Hawaii, Australia, India, New Zealand, etc. Ironic, bemused reports on peoples, customs, climate, flora and fauna, politics, much more. 197 illustrations. 720pp. 5⅜ × 8½.   26113-1 Pa. $15.95

THE PEOPLE CALLED SHAKERS, Edward D. Andrews. Definitive study of Shakers: origins, beliefs, practices, dances, social organization, furniture and crafts, etc. 33 illustrations. 351pp. 5⅜ × 8½.   21081-2 Pa. $8.95

THE MYTHS OF GREECE AND ROME, H. A. Guerber. A classic of mythology, generously illustrated, long prized for its simple, graphic, accurate retelling of the principal myths of Greece and Rome, and for its commentary on their origins and significance. With 64 illustrations by Michelangelo, Raphael, Titian, Rubens, Canova, Bernini and others. 480pp. 5⅜ × 8½.   27584-1 Pa. $9.95

PSYCHOLOGY OF MUSIC, Carl E. Seashore. Classic work discusses music as a medium from psychological viewpoint. Clear treatment of physical acoustics, auditory apparatus, sound perception, development of musical skills, nature of musical feeling, host of other topics. 88 figures. 408pp. 5⅜ × 8½.   21851-1 Pa. $9.95

THE PHILOSOPHY OF HISTORY, Georg W. Hegel. Great classic of Western thought develops concept that history is not chance but rational process, the evolution of freedom. 457pp. 5⅜ × 8½.   20112-0 Pa. $9.95

THE BOOK OF TEA, Kakuzo Okakura. Minor classic of the Orient: entertaining, charming explanation, interpretation of traditional Japanese culture in terms of tea ceremony. 94pp. 5⅜ × 8½.   20070-1 Pa. $2.95

LIFE IN ANCIENT EGYPT, Adolf Erman. Fullest, most thorough, detailed older account with much not in more recent books, domestic life, religion, magic, medicine, commerce, much more. Many illustrations reproduce tomb paintings, carvings, hieroglyphs, etc. 597pp. 5⅜ × 8½.   22632-8 Pa. $10.95

SUNDIALS, Their Theory and Construction, Albert Waugh. Far and away the best, most thorough coverage of ideas, mathematics concerned, types, construction, adjusting anywhere. Simple, nontechnical treatment allows even children to build several of these dials. Over 100 illustrations. 230pp. 5⅜ × 8½.   22947-5 Pa. $7.95

DYNAMICS OF FLUIDS IN POROUS MEDIA, Jacob Bear. For advanced students of ground water hydrology, soil mechanics and physics, drainage and irrigation engineering, and more. 335 illustrations. Exercises, with answers. 784pp. 6⅛ × 9¼.   65675-6 Pa. $19.95

SONGS OF EXPERIENCE: Facsimile Reproduction with 26 Plates in Full Color, William Blake. 26 full-color plates from a rare 1826 edition. Includes "The Tyger," "London," "Holy Thursday," and other poems. Printed text of poems. 48pp. 5¼ × 7.   24636-1 Pa. $4.95

OLD-TIME VIGNETTES IN FULL COLOR, Carol Belanger Grafton (ed.). Over 390 charming, often sentimental illustrations, selected from archives of Victorian graphics—pretty women posing, children playing, food, flowers, kittens and puppies, smiling cherubs, birds and butterflies, much more. All copyright-free. 48pp. 9¼ × 12¼.   27269-9 Pa. $5.95

PERSPECTIVE FOR ARTISTS, Rex Vicat Cole. Depth, perspective of sky and sea, shadows, much more, not usually covered. 391 diagrams, 81 reproductions of drawings and paintings. 279pp. 5⅜ × 8½. 22487-2 Pa. $6.95

DRAWING THE LIVING FIGURE, Joseph Sheppard. Innovative approach to artistic anatomy focuses on specifics of surface anatomy, rather than muscles and bones. Over 170 drawings of live models in front, back and side views, and in widely varying poses. Accompanying diagrams. 177 illustrations. Introduction. Index. 144pp. 8⅜ × 11¼. 26723-7 Pa. $7.95

GOTHIC AND OLD ENGLISH ALPHABETS: 100 Complete Fonts, Dan X. Solo. Add power, elegance to posters, signs, other graphics with 100 stunning copyright-free alphabets: Blackstone, Dolbey, Germania, 97 more—including many lower-case, numerals, punctuation marks. 104pp. 8⅛ × 11. 24695-7 Pa. $7.95

HOW TO DO BEADWORK, Mary White. Fundamental book on craft from simple projects to five-bead chains and woven works. 106 illustrations. 142pp. 5⅜ × 8. 20697-1 Pa. $4.95

THE BOOK OF WOOD CARVING, Charles Marshall Sayers. Finest book for beginners discusses fundamentals and offers 34 designs. "Absolutely first rate . . . well thought out and well executed."—E. J. Tangerman. 118pp. 7¾ × 10⅝. 23654-4 Pa. $5.95

ILLUSTRATED CATALOG OF CIVIL WAR MILITARY GOODS: Union Army Weapons, Insignia, Uniform Accessories, and Other Equipment, Schuyler, Hartley, and Graham. Rare, profusely illustrated 1846 catalog includes Union Army uniform and dress regulations, arms and ammunition, coats, insignia, flags, swords, rifles, etc. 226 illustrations. 160pp. 9 × 12. 24939-5 Pa. $10.95

WOMEN'S FASHIONS OF THE EARLY 1900s: An Unabridged Republication of "New York Fashions, 1909," National Cloak & Suit Co. Rare catalog of mail-order fashions documents women's and children's clothing styles shortly after the turn of the century. Captions offer full descriptions, prices. Invaluable resource for fashion, costume historians. Approximately 725 illustrations. 128pp. 8⅜ × 11¼. 27276-1 Pa. $11.95

THE 1912 AND 1915 GUSTAV STICKLEY FURNITURE CATALOGS, Gustav Stickley. With over 200 detailed illustrations and descriptions, these two catalogs are essential reading and reference materials and identification guides for Stickley furniture. Captions cite materials, dimensions and prices. 112pp. 6½ × 9¼. 26676-1 Pa. $9.95

EARLY AMERICAN LOCOMOTIVES, John H. White, Jr. Finest locomotive engravings from early 19th century: historical (1804–74), main-line (after 1870), special, foreign, etc. 147 plates. 142pp. 11⅜ × 8¼. 22772-3 Pa. $8.95

THE TALL SHIPS OF TODAY IN PHOTOGRAPHS, Frank O. Braynard. Lavishly illustrated tribute to nearly 100 majestic contemporary sailing vessels: Amerigo Vespucci, Clearwater, Constitution, Eagle, Mayflower, Sea Cloud, Victory, many more. Authoritative captions provide statistics, background on each ship. 190 black-and-white photographs and illustrations. Introduction. 128pp. 8⅞ × 11¾. 27163-3 Pa. $13.95

EARLY NINETEENTH-CENTURY CRAFTS AND TRADES, Peter Stockham (ed.). Extremely rare 1807 volume describes to youngsters the crafts and trades of the day: brickmaker, weaver, dressmaker, bookbinder, ropemaker, saddler, many more. Quaint prose, charming illustrations for each craft. 20 black-and-white line illustrations. 192pp. 4⅝ × 6. 27293-1 Pa. $4.95

VICTORIAN FASHIONS AND COSTUMES FROM HARPER'S BAZAR, 1867–1898, Stella Blum (ed.). Day costumes, evening wear, sports clothes, shoes, hats, other accessories in over 1,000 detailed engravings. 320pp. 9⅜ × 12¼. 22990-4 Pa. $13.95

GUSTAV STICKLEY, THE CRAFTSMAN, Mary Ann Smith. Superb study surveys broad scope of Stickley's achievement, especially in architecture. Design philosophy, rise and fall of the Craftsman empire, descriptions and floor plans for many Craftsman houses, more. 86 black-and-white halftones. 31 line illustrations. Introduction. 208pp. 6½ × 9¼. 27210-9 Pa. $9.95

THE LONG ISLAND RAIL ROAD IN EARLY PHOTOGRAPHS, Ron Ziel. Over 220 rare photos, informative text document origin (1844) and development of rail service on Long Island. Vintage views of early trains, locomotives, stations, passengers, crews, much more. Captions. 8⅞ × 11¾. 26301-0 Pa. $13.95

THE BOOK OF OLD SHIPS: From Egyptian Galleys to Clipper Ships, Henry B. Culver. Superb, authoritative history of sailing vessels, with 80 magnificent line illustrations. Galley, bark, caravel, longship, whaler, many more. Detailed, informative text on each vessel by noted naval historian. Introduction. 256pp. 5⅜ × 8½. 27332-6 Pa. $6.95

TEN BOOKS ON ARCHITECTURE, Vitruvius. The most important book ever written on architecture. Early Roman aesthetics, technology, classical orders, site selection, all other aspects. Morgan translation. 331pp. 5⅜ × 8½. 20645-9 Pa. $8.95

THE HUMAN FIGURE IN MOTION, Eadweard Muybridge. More than 4,500 stopped-action photos, in action series, showing undraped men, women, children jumping, lying down, throwing, sitting, wrestling, carrying, etc. 390pp. 7⅞ × 10⅝. 20204-6 Clothbd. $24.95

TREES OF THE EASTERN AND CENTRAL UNITED STATES AND CANADA, William M. Harlow. Best one-volume guide to 140 trees. Full descriptions, woodlore, range, etc. Over 600 illustrations. Handy size. 288pp. 4½ × 6⅜. 20395-6 Pa. $5.95

SONGS OF WESTERN BIRDS, Dr. Donald J. Borror. Complete song and call repertoire of 60 western species, including flycatchers, juncoes, cactus wrens, many more—includes fully illustrated booklet. Cassette and manual 99913-0 $8.95

GROWING AND USING HERBS AND SPICES, Milo Miloradovich. Versatile handbook provides all the information needed for cultivation and use of all the herbs and spices available in North America. 4 illustrations. Index. Glossary. 236pp. 5⅜ × 8½. 25058-X Pa. $5.95

BIG BOOK OF MAZES AND LABYRINTHS, Walter Shepherd. 50 mazes and labyrinths in all—classical, solid, ripple, and more—in one great volume. Perfect inexpensive puzzler for clever youngsters. Full solutions. 112pp. 8⅛ × 11. 22951-3 Pa. $3.95

PIANO TUNING, J. Cree Fischer. Clearest, best book for beginner, amateur. Simple repairs, raising dropped notes, tuning by easy method of flattened fifths. No previous skills needed. 4 illustrations. 201pp. 5⅜ × 8½. 23267-0 Pa. $5.95

A SOURCE BOOK IN THEATRICAL HISTORY, A. M. Nagler. Contemporary observers on acting, directing, make-up, costuming, stage props, machinery, scene design, from Ancient Greece to Chekhov. 611pp. 5⅜ × 8½. 20515-0 Pa. $11.95

THE COMPLETE NONSENSE OF EDWARD LEAR, Edward Lear. All nonsense limericks, zany alphabets, Owl and Pussycat, songs, nonsense botany, etc., illustrated by Lear. Total of 320pp. 5⅜ × 8½. (USO) 20167-8 Pa. $6.95

VICTORIAN PARLOUR POETRY: An Annotated Anthology, Michael R. Turner. 117 gems by Longfellow, Tennyson, Browning, many lesser-known poets. "The Village Blacksmith," "Curfew Must Not Ring Tonight," "Only a Baby Small," dozens more, often difficult to find elsewhere. Index of poets, titles, first lines. xxiii + 325pp. 5⅜ × 8¼. 27044-0 Pa. $8.95

DUBLINERS, James Joyce. Fifteen stories offer vivid, tightly focused observations of the lives of Dublin's poorer classes. At least one, "The Dead," is considered a masterpiece. Reprinted complete and unabridged from standard edition. 160pp. 5³⁄₁₆ × 8¼. 26870-5 Pa. $1.00

THE HAUNTED MONASTERY and THE CHINESE MAZE MURDERS, Robert van Gulik. Two full novels by van Gulik, set in 7th-century China, continue adventures of Judge Dee and his companions. An evil Taoist monastery, seemingly supernatural events; overgrown topiary maze hides strange crimes. 27 illustrations. 328pp. 5⅜ × 8½. 23502-5 Pa. $7.95

THE BOOK OF THE SACRED MAGIC OF ABRAMELIN THE MAGE, translated by S. MacGregor Mathers. Medieval manuscript of ceremonial magic. Basic document in Aleister Crowley, Golden Dawn groups. 268pp. 5⅜ × 8½. 23211-5 Pa. $8.95

NEW RUSSIAN-ENGLISH AND ENGLISH-RUSSIAN DICTIONARY, M. A. O'Brien. This is a remarkably handy Russian dictionary, containing a surprising amount of information, including over 70,000 entries. 366pp. 4½ × 6⅛. 20208-9 Pa. $9.95

HISTORIC HOMES OF THE AMERICAN PRESIDENTS, Second, Revised Edition, Irvin Haas. A traveler's guide to American Presidential homes, most open to the public, depicting and describing homes occupied by every American President from George Washington to George Bush. With visiting hours, admission charges, travel routes. 175 photographs. Index. 160pp. 8¼ × 11. 26751-2 Pa. $10.95

NEW YORK IN THE FORTIES, Andreas Feininger. 162 brilliant photographs by the well-known photographer, formerly with *Life* magazine. Commuters, shoppers, Times Square at night, much else from city at its peak. Captions by John von Hartz. 181pp. 9¼ × 10¾. 23585-8 Pa. $12.95

INDIAN SIGN LANGUAGE, William Tomkins. Over 525 signs developed by Sioux and other tribes. Written instructions and diagrams. Also 290 pictographs. 111pp. 6⅛ × 9¼. 22029-X Pa. $3.50

ANATOMY: A Complete Guide for Artists, Joseph Sheppard. A master of figure drawing shows artists how to render human anatomy convincingly. Over 460 illustrations. 224pp. 8⅜ × 11¼. 27279-6 Pa. $9.95

MEDIEVAL CALLIGRAPHY: Its History and Technique, Marc Drogin. Spirited history, comprehensive instruction manual covers 13 styles (ca. 4th century thru 15th). Excellent photographs; directions for duplicating medieval techniques with modern tools. 224pp. 8⅜ × 11¼. 26142-5 Pa. $11.95

DRIED FLOWERS: How to Prepare Them, Sarah Whitlock and Martha Rankin. Complete instructions on how to use silica gel, meal and borax, perlite aggregate, sand and borax, glycerine and water to create attractive permanent flower arrangements. 12 illustrations. 32pp. 5⅜ × 8½. 21802-3 Pa. $1.00

EASY-TO-MAKE BIRD FEEDERS FOR WOODWORKERS, Scott D. Campbell. Detailed, simple-to-use guide for designing, constructing, caring for and using feeders. Text, illustrations for 12 classic and contemporary designs. 96pp. 5⅜ × 8½. 25847-5 Pa. $2.95

OLD-TIME CRAFTS AND TRADES, Peter Stockham. An 1807 book created to teach children about crafts and trades open to them as future careers. It describes in detailed, nontechnical terms 24 different occupations, among them coachmaker, gardener, hairdresser, lacemaker, shoemaker, wheelwright, copper-plate printer, milliner, trunkmaker, merchant and brewer. Finely detailed engravings illustrate each occupation. 192pp. 4⅝ × 6. 27398-9 Pa. $4.95

THE HISTORY OF UNDERCLOTHES, C. Willett Cunnington and Phyllis Cunnington. Fascinating, well-documented survey covering six centuries of English undergarments, enhanced with over 100 illustrations: 12th-century laced-up bodice, footed long drawers (1795), 19th-century bustles, 19th-century corsets for men, Victorian "bust improvers," much more. 272pp. 5⅜ × 8¼. 27124-2 Pa. $9.95

ARTS AND CRAFTS FURNITURE: The Complete Brooks Catalog of 1912, Brooks Manufacturing Co. Photos and detailed descriptions of more than 150 now very collectible furniture designs from the Arts and Crafts movement depict davenports, settees, buffets, desks, tables, chairs, bedsteads, dressers and more, all built of solid, quarter-sawed oak. Invaluable for students and enthusiasts of antiques, Americana and the decorative arts. 80pp. 6½ × 9¼. 27471-3 Pa. $7.95

HOW WE INVENTED THE AIRPLANE: An Illustrated History, Orville Wright. Fascinating firsthand account covers early experiments, construction of planes and motors, first flights, much more. Introduction and commentary by Fred C. Kelly. 76 photographs. 96pp. 8¼ × 11. 25662-6 Pa. $8.95

THE ARTS OF THE SAILOR: Knotting, Splicing and Ropework, Hervey Garrett Smith. Indispensable shipboard reference covers tools, basic knots and useful hitches; handsewing and canvas work, more. Over 100 illustrations. Delightful reading for sea lovers. 256pp. 5⅜ × 8½. 26440-8 Pa. $7.95

FRANK LLOYD WRIGHT'S FALLINGWATER: The House and Its History, Second, Revised Edition, Donald Hoffmann. A total revision—both in text and illustrations—of the standard document on Fallingwater, the boldest, most personal architectural statement of Wright's mature years, updated with valuable new material from the recently opened Frank Lloyd Wright Archives. "Fascinating"—*The New York Times*. 116 illustrations. 128pp. 9¼ × 10¾. 27430-6 Pa. $10.95

PHOTOGRAPHIC SKETCHBOOK OF THE CIVIL WAR, Alexander Gardner. 100 photos taken on field during the Civil War. Famous shots of Manassas, Harper's Ferry, Lincoln, Richmond, slave pens, etc. 244pp. 10⅝ × 8¼.
22731-6 Pa. $9.95

FIVE ACRES AND INDEPENDENCE, Maurice G. Kains. Great back-to-the-land classic explains basics of self-sufficient farming. The one book to get. 95 illustrations. 397pp. 5⅜ × 8½.   20974-1 Pa. $7.95

SONGS OF EASTERN BIRDS, Dr. Donald J. Borror. Songs and calls of 60 species most common to eastern U.S.: warblers, woodpeckers, flycatchers, thrushes, larks, many more in high-quality recording.   Cassette and manual 99912-2 $8.95

A MODERN HERBAL, Margaret Grieve. Much the fullest, most exact, most useful compilation of herbal material. Gigantic alphabetical encyclopedia, from aconite to zedoary, gives botanical information, medical properties, folklore, economic uses, much else. Indispensable to serious reader. 161 illustrations. 888pp. 6½ × 9¼. 2-vol. set. (USO)   Vol. I: 22798-7 Pa. $9.95
Vol. II: 22799-5 Pa. $9.95

HIDDEN TREASURE MAZE BOOK, Dave Phillips. Solve 34 challenging mazes accompanied by heroic tales of adventure. Evil dragons, people-eating plants, bloodthirsty giants, many more dangerous adversaries lurk at every twist and turn. 34 mazes, stories, solutions. 48pp. 8¼ × 11.   24566-7 Pa. $2.95

LETTERS OF W. A. MOZART, Wolfgang A. Mozart. Remarkable letters show bawdy wit, humor, imagination, musical insights, contemporary musical world; includes some letters from Leopold Mozart. 276pp. 5⅜ × 8½.   22859-2 Pa. $6.95

BASIC PRINCIPLES OF CLASSICAL BALLET, Agrippina Vaganova. Great Russian theoretician, teacher explains methods for teaching classical ballet. 118 illustrations. 175pp. 5⅜ × 8½.   22036-2 Pa. $4.95

THE JUMPING FROG, Mark Twain. Revenge edition. The original story of The Celebrated Jumping Frog of Calaveras County, a hapless French translation, and Twain's hilarious "retranslation" from the French. 12 illustrations. 66pp. 5⅜ × 8½.
22686-7 Pa. $3.95

BEST REMEMBERED POEMS, Martin Gardner (ed.). The 126 poems in this superb collection of 19th- and 20th-century British and American verse range from Shelley's "To a Skylark" to the impassioned "Renascence" of Edna St. Vincent Millay and to Edward Lear's whimsical "The Owl and the Pussycat." 224pp. 5⅜ × 8½.
27165-X Pa. $4.95

COMPLETE SONNETS, William Shakespeare. Over 150 exquisite poems deal with love, friendship, the tyranny of time, beauty's evanescence, death and other themes in language of remarkable power, precision and beauty. Glossary of archaic terms. 80pp. 5³⁄₁₆ × 8¼.   26686-9 Pa. $1.00

BODIES IN A BOOKSHOP, R. T. Campbell. Challenging mystery of blackmail and murder with ingenious plot and superbly drawn characters. In the best tradition of British suspense fiction. 192pp. 5⅜ × 8½.   24720-1 Pa. $5.95

CATALOG OF DOVER BOOKS

THE WIT AND HUMOR OF OSCAR WILDE, Alvin Redman (ed.). More than 1,000 ripostes, paradoxes, wisecracks: Work is the curse of the drinking classes; I can resist everything except temptation; etc. 258pp. 5⅜ × 8½.   20602-5 Pa. $5.95

SHAKESPEARE LEXICON AND QUOTATION DICTIONARY, Alexander Schmidt. Full definitions, locations, shades of meaning in every word in plays and poems. More than 50,000 exact quotations. 1,485pp. 6½ × 9¼. 2-vol. set.
Vol. 1: 22726-X Pa. $15.95
Vol. 2: 22727-8 Pa. $15.95

SELECTED POEMS, Emily Dickinson. Over 100 best-known, best-loved poems by one of America's foremost poets, reprinted from authoritative early editions. No comparable edition at this price. Index of first lines. 64pp. 5³⁄₁₆ × 8¼.
26466-1 Pa. $1.00

CELEBRATED CASES OF JUDGE DEE (DEE GOONG AN), translated by Robert van Gulik. Authentic 18th-century Chinese detective novel; Dee and associates solve three interlocked cases. Led to van Gulik's own stories with same characters. Extensive introduction. 9 illustrations. 237pp. 5⅜ × 8½.
23337-5 Pa. $6.95

THE MALLEUS MALEFICARUM OF KRAMER AND SPRENGER, translated by Montague Summers. Full text of most important witchhunter's "bible," used by both Catholics and Protestants. 278pp. 6⅝ × 10.   22802-9 Pa. $10.95

SPANISH STORIES/CUENTOS ESPAÑOLES: A Dual-Language Book, Angel Flores (ed.). Unique format offers 13 great stories in Spanish by Cervantes, Borges, others. Faithful English translations on facing pages. 352pp. 5⅜ × 8½.
25399-6 Pa. $8.95

THE CHICAGO WORLD'S FAIR OF 1893: A Photographic Record, Stanley Appelbaum (ed.). 128 rare photos show 200 buildings, Beaux-Arts architecture, Midway, original Ferris Wheel, Edison's kinetoscope, more. Architectural emphasis; full text. 116pp. 8¼ × 11.   23990-X Pa. $9.95

OLD QUEENS, N.Y., IN EARLY PHOTOGRAPHS, Vincent F. Seyfried and William Asadorian. Over 160 rare photographs of Maspeth, Jamaica, Jackson Heights, and other areas. Vintage views of DeWitt Clinton mansion, 1939 World's Fair and more. Captions. 192pp. 8⅞ × 11.   26358-4 Pa. $12.95

CAPTURED BY THE INDIANS: 15 Firsthand Accounts, 1750–1870, Frederick Drimmer. Astounding true historical accounts of grisly torture, bloody conflicts, relentless pursuits, miraculous escapes and more, by people who lived to tell the tale. 384pp. 5⅜ × 8½.   24901-8 Pa. $8.95

THE WORLD'S GREAT SPEECHES, Lewis Copeland and Lawrence W. Lamm (eds.). Vast collection of 278 speeches of Greeks to 1970. Powerful and effective models; unique look at history. 842pp. 5⅜ × 8½.   20468-5 Pa. $13.95

THE BOOK OF THE SWORD, Sir Richard F. Burton. Great Victorian scholar/adventurer's eloquent, erudite history of the "queen of weapons"—from prehistory to early Roman Empire. Evolution and development of early swords, variations (sabre, broadsword, cutlass, scimitar, etc.), much more. 336pp. 6⅛ × 9¼. 25434-8 Pa. $8.95

AUTOBIOGRAPHY: The Story of My Experiments with Truth, Mohandas K. Gandhi. Boyhood, legal studies, purification, the growth of the Satyagraha (nonviolent protest) movement. Critical, inspiring work of the man responsible for the freedom of India. 480pp. 5⅜ × 8½. (USO) 24593-4 Pa. $7.95

CELTIC MYTHS AND LEGENDS, T. W. Rolleston. Masterful retelling of Irish and Welsh stories and tales. Cuchulain, King Arthur, Deirdre, the Grail, many more. First paperback edition. 58 full-page illustrations. 512pp. 5⅜ × 8½. 26507-2 Pa. $9.95

THE PRINCIPLES OF PSYCHOLOGY, William James. Famous long course complete, unabridged. Stream of thought, time perception, memory, experimental methods; great work decades ahead of its time. 94 figures. 1,391pp. 5⅜ × 8½. 2-vol. set.
Vol. I: 20381-6 Pa. $12.95
Vol. II: 20382-4 Pa. $12.95

THE WORLD AS WILL AND REPRESENTATION, Arthur Schopenhauer. Definitive English translation of Schopenhauer's life work, correcting more than 1,000 errors, omissions in earlier translations. Translated by E. F. J. Payne. Total of 1,269pp. 5⅜ × 8½. 2-vol. set.
Vol. 1: 21761-2 Pa. $11.95
Vol. 2: 21762-0 Pa. $11.95

MAGIC AND MYSTERY IN TIBET, Madame Alexandra David-Neel. Experiences among lamas, magicians, sages, sorcerers, Bonpa wizards. A true psychic discovery. 32 illustrations. 321pp. 5⅜ × 8½. (USO) 22682-4 Pa. $8.95

THE EGYPTIAN BOOK OF THE DEAD, E. A. Wallis Budge. Complete reproduction of Ani's papyrus, finest ever found. Full hieroglyphic text, interlinear transliteration, word-for-word translation, smooth translation. 533pp. 6½ × 9¼. 21866-X Pa. $9.95

MATHEMATICS FOR THE NONMATHEMATICIAN, Morris Kline. Detailed, college-level treatment of mathematics in cultural and historical context, with numerous exercises. Recommended Reading Lists. Tables. Numerous figures. 641pp. 5⅜ × 8½. 24823-2 Pa. $11.95

THEORY OF WING SECTIONS: Including a Summary of Airfoil Data, Ira H. Abbott and A. E. von Doenhoff. Concise compilation of subsonic aerodynamic characteristics of NACA wing sections, plus description of theory. 350pp. of tables. 693pp. 5⅜ × 8½. 60586-8 Pa. $13.95

THE RIME OF THE ANCIENT MARINER, Gustave Doré, S. T. Coleridge. Doré's finest work; 34 plates capture moods, subtleties of poem. Flawless full-size reproductions printed on facing pages with authoritative text of poem. "Beautiful. Simply beautiful."—*Publisher's Weekly*. 77pp. 9¼ × 12. 22305-1 Pa. $5.95

NORTH AMERICAN INDIAN DESIGNS FOR ARTISTS AND CRAFTSPEOPLE, Eva Wilson. Over 360 authentic copyright-free designs adapted from Navajo blankets, Hopi pottery, Sioux buffalo hides, more. Geometrics, symbolic figures, plant and animal motifs, etc. 128pp. 8⅜ × 11. (EUK) 25341-4 Pa. $7.95

SCULPTURE: Principles and Practice, Louis Slobodkin. Step-by-step approach to clay, plaster, metals, stone; classical and modern. 253 drawings, photos. 255pp. 8⅛ × 11. 22960-2 Pa. $10.95

THE INFLUENCE OF SEA POWER UPON HISTORY, 1660–1783, A. T. Mahan. Influential classic of naval history and tactics still used as text in war colleges. First paperback edition. 4 maps. 24 battle plans. 640pp. 5⅜ × 8½.
25509-3 Pa. $12.95

THE STORY OF THE TITANIC AS TOLD BY ITS SURVIVORS, Jack Winocour (ed.). What it was really like. Panic, despair, shocking inefficiency, and a little heroism. More thrilling than any fictional account. 26 illustrations. 320pp. 5⅜ × 8½.
20610-6 Pa. $7.95

FAIRY AND FOLK TALES OF THE IRISH PEASANTRY, William Butler Yeats (ed.). Treasury of 64 tales from the twilight world of Celtic myth and legend: "The Soul Cages," "The Kildare Pooka," "King O'Toole and his Goose," many more. Introduction and Notes by W. B. Yeats. 352pp. 5⅜ × 8½. 26941-8 Pa. $8.95

BUDDHIST MAHAYANA TEXTS, E. B. Cowell and Others (eds.). Superb, accurate translations of basic documents in Mahayana Buddhism, highly important in history of religions. The Buddha-karita of Asvaghosha, Larger Sukhavativyuha, more. 448pp. 5⅜ × 8½. , 25552-2 Pa. $9.95

ONE TWO THREE . . . INFINITY: Facts and Speculations of Science, George Gamow. Great physicist's fascinating, readable overview of contemporary science: number theory, relativity, fourth dimension, entropy, genes, atomic structure, much more. 128 illustrations. Index. 352pp. 5⅜ × 8½. 25664-2 Pa. $8.95

ENGINEERING IN HISTORY, Richard Shelton Kirby, et al. Broad, nontechnical survey of history's major technological advances: birth of Greek science, industrial revolution, electricity and applied science, 20th-century automation, much more. 181 illustrations. ". . . excellent . . ."—Isis. Bibliography. vii + 530pp. 5⅜ × 8¼.
26412-2 Pa. $14.95

Prices subject to change without notice.

Available at your book dealer or write for free catalog to Dept. GI, Dover Publications, Inc., 31 East 2nd St., Mineola, N.Y. 11501. Dover publishes more than 500 books each year on science, elementary and advanced mathematics, biology, music, art, literary history, social sciences and other areas.